KUMON MATH WORKBOOKS

Grade **3**

Word Problems

Table of Contents

KUMON

1 Claire and James played "I Spy" in the car. Claire won 46 times, and James won 7 times. How many more times did Claire win than James?

10 points

⟨Ans.⟩ _____

2 The cafeteria had 65 jugs of milk. 35 jugs were sold. How many jugs of milk does the cafeteria have remaining?

10 points

⟨Ans.⟩ _____

3 We bought a blanket for grandmother that is 2 meters 35 centimeters long and 1 meter 70 centimeters wide. What is the difference between the length and width of grandmother's blanket?

10 points

⟨Ans.⟩ _____

4 There were 20 candies in the bowl by the door this morning. Malik ate some candies while he was waiting for the doctor, and now there are 12 candies. How many candies did Malik eat?

10 points

⟨Ans.⟩ _____

5 Your mother gave you 40 stickers for your birthday. You gave 7 of them to your brother and 9 of them to your sister. How many stickers do you have left?

10 points

⟨Ans.⟩ _____

6 Today, Ted read 45 pages for class. Yesterday and today, he read 93 pages in all. How many pages did Ted read yesterday?

10 points

⟨Ans.⟩ _____

7 Anne is looking at a 90¢ notebook at the store. The price of the notebook is 40¢ higher than the price of an eraser. How much is the eraser?

10 points

⟨Ans.⟩ _____

8 Those pigeons are in our tree again. 4 pigeons just flew away, and 6 more landed. Now there are 35 total pigeons in our tree. How many were there at the beginning?

10 points

⟨Ans.⟩ _____

9 The teacher has 25 candies. If he wants to give one candy to each of us, and there are 26 of us in class, how many more candies does the teacher need?

10 points

⟨Ans.⟩ _____

10 The sailor has a rope that is 10 feet 10 inches long, and another one that is 15 feet 1 inch long. If he connects the two ropes, how long will his new rope be?

10 points

⟨Ans.⟩ _____

Do you remember this? Good!

3

1 The school has 65 notebooks. If they want to give 1 notebook to each child, and there are 90 children in the school, how many more notebooks does the school need?

10 points

〈Ans.〉

2 Kim has 32 flowers in her garden. Today, she cut 6 flowers. How many flowers remain?

10 points

〈Ans.〉

3 Sara is 4 feet 5 inches tall. When she stands on the 6-inch step, how tall is she?

10 points

〈Ans.〉

4 Jack solved 28 problems for homework, but there are still 22 problems remaining. How many problems was Jack assigned in all?

10 points

〈Ans.〉

5 You bought a snack during recess and paid 57¢. If you have 13¢ left, how much money did you have at first?

10 points

〈Ans.〉

6 In front of the school, there is a line for the bus. 18 people joined the line, and now there are 38 people in line. How many people were waiting for the bus at first?

10 points

⟨Ans.⟩

7 There are 95 children playing in the park today. If there are 46 more children than adults, how many adults are in the park today?

10 points

⟨Ans.⟩

8 Yesterday, there were 20 bags of sugar in the warehouse. Then you brought 32 more bags, and today you brought 14 more bags. How many more bags are in the warehouse now?

10 points

⟨Ans.⟩

9 Sue's flowerbed is 5 meters long. The width is 10 meters more than the length. How wide is her flowerbed?

10 points

⟨Ans.⟩

10 You have colored sheets of paper for crafts class. Today, you got 25 more from your teacher and used 10. Now, you have 18 sheets of colored paper left. How many sheets did you have before class today?

10 points

⟨Ans.⟩

Now let's get going!

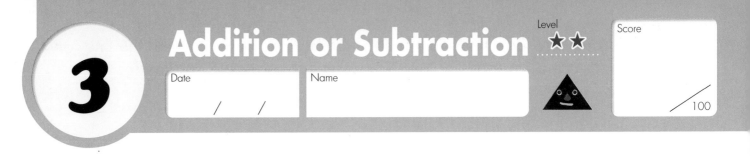

Addition or Subtraction

Date / /

Name

Score /100

1 Sam was thirsty this morning and drank 240 milliliters of juice and 250 milliliters of cola. How much liquid did he drink in all?

10 points

⟨Ans.⟩

2 In the pantry, you had 450 grams of sugar. Today, your mother added 85 grams of sugar. How much sugar is in there now?

10 points

⟨Ans.⟩

3 While waiting in line at the amusement park, Freddie counted the flowers. He counted 135 red flowers and 122 white flowers. How many flowers did he count in all?

10 points

⟨Ans.⟩

4 There are two rolls of tape in class. One is 450 centimeters long, and the other is 680 centimeters long. How much tape do we have altogether?

10 points

⟨Ans.⟩

5 Ken's school has 178 boys and 169 girls. How many students are there in all?

10 points

⟨Ans.⟩

6 Yesterday, 245 adults came to the movie theater. If 178 more children than adults came to the movie theater, how many children came to the movie theater yesterday?

10 points

⟨**Ans.**⟩ _____

7 At the book fair, the librarians bought 283 storybooks and 175 picture books. How many books did they buy in all?

10 points

⟨**Ans.**⟩ _____

8 Tammy's class folded 359 paper cranes yesterday, and 486 today. How many cranes did they fold altogether?

10 points

⟨**Ans.**⟩ _____

9 Ben's class went out picking strawberries. Ben found 128 strawberries while Cathy found 147. How many strawberries did Ben and Cathy pick altogether?

10 points

⟨**Ans.**⟩ _____

10 Lisa and Dennis wanted to clean up their street. Lisa picked up 165 pieces of trash, and Dennis picked up 187. How many pieces of trash did they pick up in all?

10 points

⟨**Ans.**⟩ _____

You're doing really well!

7

4 Addition or Subtraction

Date   /   /

Name

Score   /100

1   Ben had 770 milliliters of juice. If he drank 250 milliliters of juice this morning, how much juice does he have remaining?   10 points

⟨Ans.⟩ _____

2   In the flowerbed in front of school, there are 185 red flowers and 133 yellow ones. How many more red flowers are there?   10 points

⟨Ans.⟩ _____

3   May had 128 stickers, and she gave 75 to her sister. How many does she have now?   10 points

⟨Ans.⟩ _____

4   There are 178 boys and 169 girls at Andy's school. Are there more boys or girls? How many more?   10 points

⟨Ans.⟩ There are _____ more _____.

5   The gym teacher told Rebecca to run 800 yards. She has run 245 yards so far. How many more yards does she still have to run?   10 points

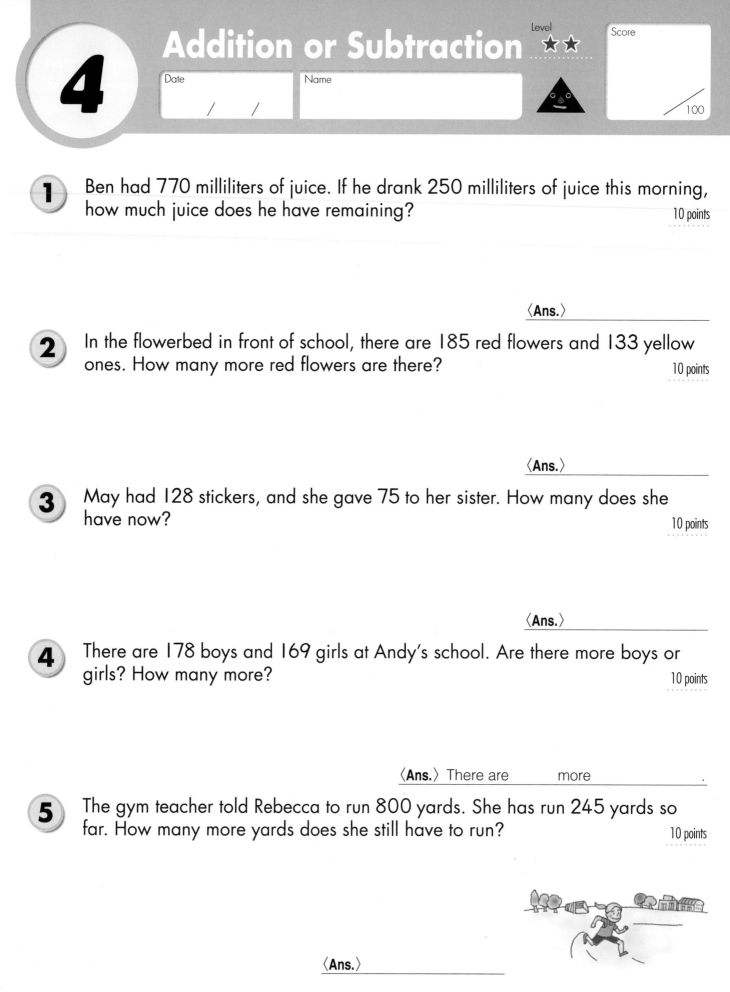

⟨Ans.⟩ _____

6 Prudence read 83 pages in her book. If her book is 240 pages long, how many pages does she have left?

10 points

〈Ans.〉 _____

7 There are 564 students at Brandon's school. If 296 are boys, how many girls are at his school?

10 points

〈Ans.〉 _____

8 Bob is helping Mrs. Chen find her lost cat. He made 265 copies of her poster. The copying machine had 500 sheets of paper when he began. How many sheets of paper are left in the machine?

10 points

〈Ans.〉 _____

9 Of the 562 students at Bill's school, 175 had cavities. How many students had no cavities?

10 points

〈Ans.〉 _____

10 On Tuesday, 324 children came to the amusement park. If 143 less adults than children came to the park, how many adults came to the park?

10 points

〈Ans.〉 _____

The numbers are getting bigger, but you're doing fine.

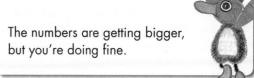

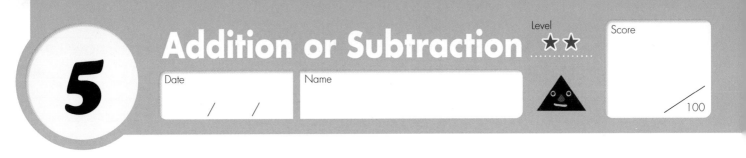

1 264 people were in the amusement park. It started raining, so 86 people went home. How many people were left?   10 points

⟨Ans.⟩

2 Kevin's school has 175 boys and 179 girls. How many people are there altogether in Kevin's school?   10 points

⟨Ans.⟩

3 548 people came to the art exhibition yesterday, and 497 people came today. How many people came to the art exhibition in all?   10 points

⟨Ans.⟩

4 Ava saved up 500 pennies. She wanted some candy, so she used 348 of the pennies. How many pennies does she have left?   10 points

⟨Ans.⟩

5 Rima has 125 stickers and her sister has 175. How many stickers do they have altogether?   10 points

⟨Ans.⟩

6 Nick's book about pirates is exciting, and he read 176 pages yesterday. If the book has 240 pages, how many pages are remaining?

10 points

⟨Ans.⟩ _____

7 Terry's school has 365 students. There are 196 boys. How many girls are there?

10 points

⟨Ans.⟩ _____

8 Captain Ned has a 500-inch rope and a 265-inch rope. How much rope does Captain Ned have altogether?

10 points

⟨Ans.⟩ _____

9 The zookeeper has to feed a panda and her cub. Then panda will eat 360 bamboo sticks, and the cub will eat 250 sticks. How many sticks does the zookeeper need?

10 points

⟨Ans.⟩ _____

10 The cafeteria had 263 oranges this morning. Children ate 185 of them today. How many oranges does the cafeteria have left?

10 points

⟨Ans.⟩ _____

Let's try something new!

1 The distance from Ella's house to her grandparents' house is 300 miles. From her grandparents' house to the state capital is 500 miles. How far is it from Ella's house to the capital if they stop at her grandparents' house along the way?

10 points

$$300 \text{ mi.} + 500 \text{ mi.} =$$

⟨Ans.⟩ _____

2 From Michele's house to the park, it is 400 feet. From the park to the library, it is 200 feet. How far is it from her house to the library if she passes through the park?

10 points

⟨Ans.⟩ _____

3 It is 500 meters from John's house to the school. From John's house to the bookstore, it is 300 meters. If John has to go pick up a book from the bookstore on his way home after school, how far is it from the school to the bookstore?

10 points

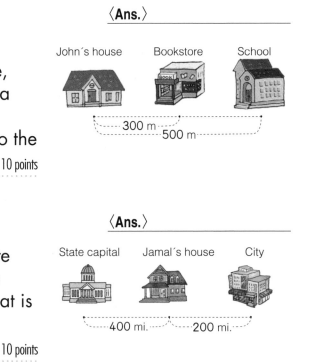

John's house   Bookstore   School

300 m
500 m

⟨Ans.⟩ _____

4 The distance from Jamal's house to the state capital is 400 miles. The distance from his house to the nearest city is 200 miles. What is the difference between the two distances?

10 points

State capital   Jamal's house   City

400 mi.   200 mi.

⟨Ans.⟩ _____

5 The map below shows the area around Carol's house.

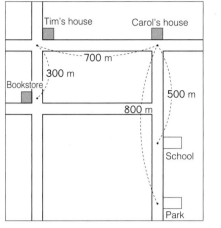

(1) If Carol passes Tim's house, how far is it from her house to the bookstore?

〈**Ans.**〉 _____

(2) How far is it from the school to the park?

〈**Ans.**〉 _____

6 Billy is hiking his favorite trail. He has already gone 4 kilometers 500 meters. He still has 1 kilometer 200 meters to reach his goal. How long is the hike altogether?

10 points

〈**Ans.**〉 _____

7 The distance from Sally's house to her uncle's house is 800 meters. Today, she walked to and from her uncle's house. How far did she walk in all?

15 points

〈**Ans.**〉 _____

8 Julian is going to the zoo, which is 2 kilometers away from his house on foot. He has already walked 1 kilometer 200 meters. How much further does he have to walk?

15 points

$$2 \text{ km} - 1 \text{ km } 200 \text{ m} =$$

〈**Ans.**〉 _____

Wow, this is tough. Good job!

13

7

Time

Level
★★

Date / /

Name

Score
/100

1 Maria left her house at 8 o'clock this morning and arrived at school at 8:16. How long did it take her to walk to school?

10 points

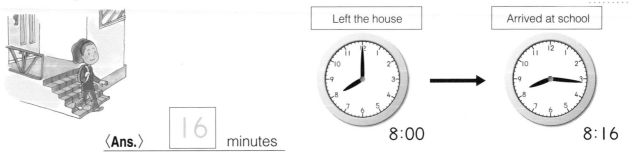

Left the house	Arrived at school
8:00	8:16

⟨Ans.⟩ 16 minutes

2 Ted left his house at 9 o'clock this morning and arrived at the train station at 9:28. How long did he take?

10 points

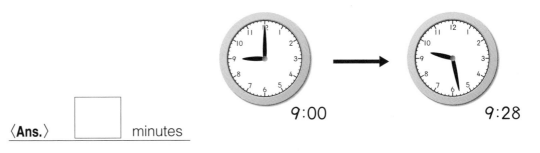

9:00   9:28

⟨Ans.⟩ [ ] minutes

3 Rosa jumped rope from 10:30 to 11:00. How long did she play?

10 points

⟨Ans.⟩ _____ minutes

4 Mike read his favorite book from 2:25 to 2:50. How long did he read?

10 points

⟨Ans.⟩ _____

5 Tim walked with his dog from 6:15 to 6:55. How long did he take?

10 points

⟨Ans.⟩ _____

6 Jim was playing in the park from 4:00 in the afternoon until 5:00. How long was he in the park?

10 points

⟨Ans.⟩ ☐ 1 hour

Start   End

4:00   5:00

7 Kim worked on a drawing from 9:00 to 11:00. How long did she spend drawing?

10 points

⟨Ans.⟩ ☐ hours

9:00   11:00

8 Mike went shopping with his mother. They left home at 9 o'clock, and when they returned it was 12 o'clock. How long did the shopping take?

10 points

⟨Ans.⟩ _____ hours

9 Sally left home at 8 o'clock in the morning to watch a movie in the library. If she came home at noon, how long was she gone?

10 points

⟨Ans.⟩ _____

10 Will's class went on a field trip, and their lunch was from 12 to 1 o'clock. How long was their lunchtime?

10 points

⟨Ans.⟩ _____

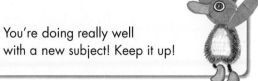

You're doing really well with a new subject! Keep it up!

8 Time

Level ★★

Date / /

Name

Score
/100

1 Mike left his house at 8 o'clock this morning, and arrived at school 15 minutes later. What time did he arrive at school?

10 points

⟨Ans.⟩ 8 : 15

2 Julia started studying math at 4 o'clock, and finished 45 minutes later. What time did she finish?

10 points

⟨Ans.⟩ [ ] : [ ]

3 Ted started jumping rope at 3:20, and finished 30 minutes later. What time did he finish?

10 points

⟨Ans.⟩ :

4 Kim started cleaning her room at 10:40, and finished 15 minutes later. What time did she finish?

10 points

⟨Ans.⟩

5 Jim's father started walking at 9:20, and finished 40 minutes later. What time did he finish?

10 points

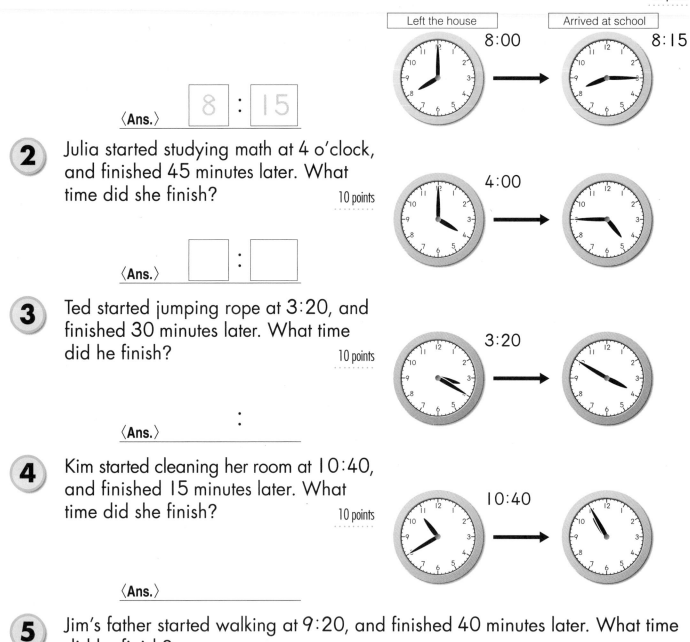

Left the house   8:00   Arrived at school   8:15

4:00

3:20

10:40

⟨Ans.⟩

6 The distance from Ted's house to the station is 10 minutes on foot. He'd like to arrive at the station at 9:00. What time does he have to leave his house?

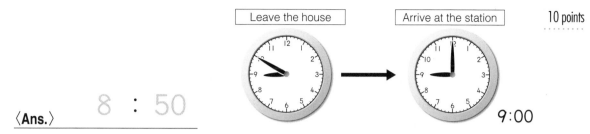

Leave the house   Arrive at the station   10 points

9:00

⟨Ans.⟩ 8 : 50

7 The distance from Bob's house to the library is 15 minutes on foot. He'd like to arrive at the library at 10:00. What time does he have to leave his house?

10 points

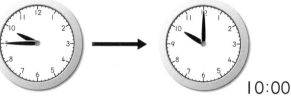

10:00

⟨Ans.⟩ :

8 Anne has been reading her book for 30 minutes. If it is 4:50 now, what time did she start?

10 points

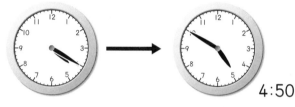

4:50

⟨Ans.⟩ :

9 Cathy has to practice her piano for 30 minutes today. If she wants to finish at 4:30, what time does she have to start?

10 points

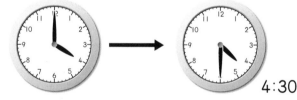

4:30

⟨Ans.⟩ :

10 Jimmy was roller-skating for 20 minutes this afternoon, and finished at 3:30. What time did he start?

10 points

⟨Ans.⟩ :

Very good. Now let's try something new!

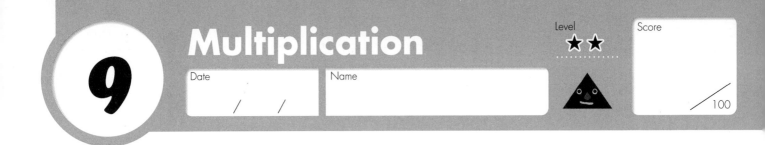

Multiplication

9

Level ★★

Date   /   /

Name

Score

/100

1 Mother made us lunch. She put 2 oranges on each dish, and there are 4 dishes on the table. How many total oranges did she serve?

10 points

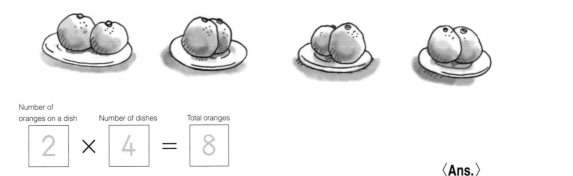

Number of oranges on a dish | Number of dishes | Total oranges

$$\boxed{2} \times \boxed{4} = \boxed{8}$$

⟨Ans.⟩ _____

2 Class is about to start. Everyone sits down. If 2 children sit on each bench, and there are 8 benches, how many children are in our class?

10 points

Number of children on each bench | Number of benches | Total children

$$\boxed{2} \times \boxed{8} = \boxed{}$$

⟨Ans.⟩ _____

3 In our living room, we have 4 vases, and each vase has 3 flowers. How many flowers are in our living room altogether?

10 points

Number of flowers in each vase | Number of vases | Total flowers

$$\boxed{3} \times \boxed{} = \boxed{}$$

⟨Ans.⟩ _____

4 Jamal's friends are all on paddleboats in the pond. There are 5 boats and 3 people on each boat. How many people are in the boats in all?

10 points

Number of people in each boat | Number of boats | Total people

$$\boxed{} \times \boxed{} = \boxed{}$$

⟨Ans.⟩ _____

5 The teacher wants to give everyone 2 pencils each. There are 9 people in the class. How many pencils will the teacher need?

10 points

☐ × ☐ = ☐

⟨Ans.⟩ _____

6 The florist is putting together some vases full of flowers for an important order. Each vase will have 3 flowers, and there are 7 vases in all. How many flowers does the florist need?

10 points

$3 \times 7 =$

⟨Ans.⟩ _____

7 In Shelly's homework group, there are 3 teams, and each team has 2 members. How many people are in Shelly's homework group?

10 points

⟨Ans.⟩ _____

8 Ally is trying to fix her notebook. She has 2 pieces of tape that are 3 centimeters each. How much tape does she have in all?

10 points

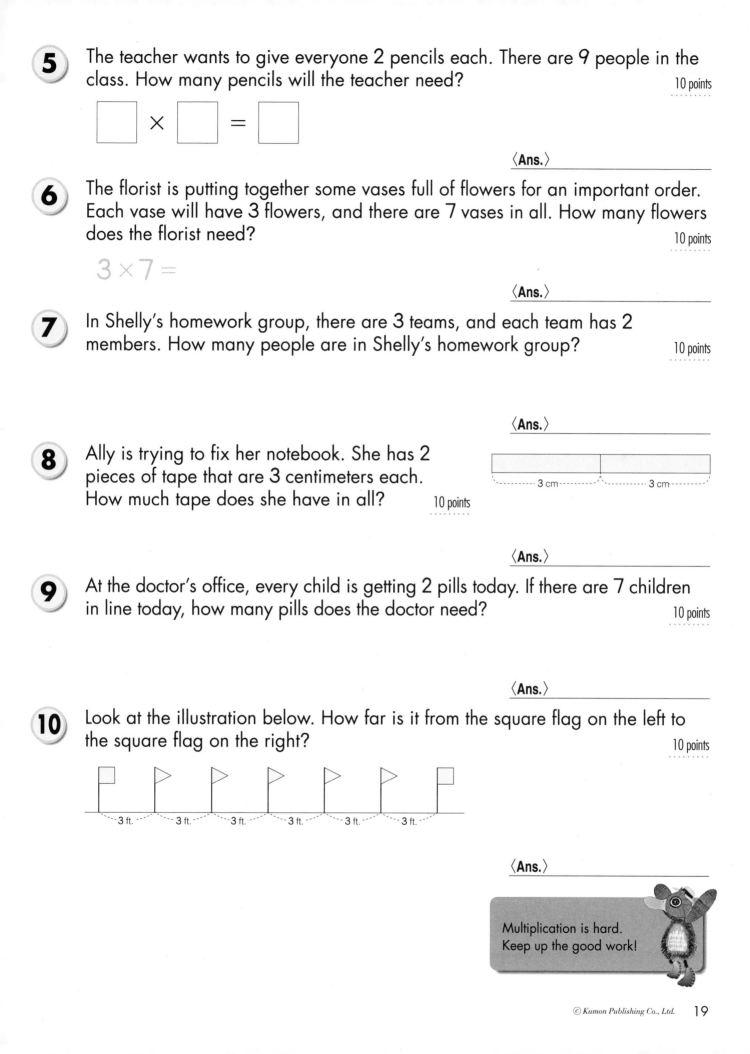

⟨Ans.⟩ _____

9 At the doctor's office, every child is getting 2 pills today. If there are 7 children in line today, how many pills does the doctor need?

10 points

⟨Ans.⟩ _____

10 Look at the illustration below. How far is it from the square flag on the left to the square flag on the right?

10 points

⟨Ans.⟩ _____

Multiplication is hard.
Keep up the good work!

10 Multiplication

Date   /   /

Name

Level ★ ★

Score

/100

1 In the library, there are 6 sofas, and 4 people can sit on each sofa. How many people can sit on the sofas in all?

10 points

Number of people per sofa   Number of sofas   Total people

☐ × ☐ = ☐

⟨Ans.⟩ _____

2 We came back from the grocery store holding 7 bags with 5 oranges in each bag. How many oranges did we buy altogether?

10 points

Number of oranges per bag   Number of bags   Total oranges

☐ × ☐ = ☐

⟨Ans.⟩ _____

3 At Monica's birthday party, her mother gave 4 candies to each child. If 5 children came to the party, how many candies did Monica's mother need?

10 points

Number of candies per child   Number of children   Total candies

☐ × ☐ = ☐

⟨Ans.⟩ _____

4 The crafts teacher gives 4 colored sheets of paper to each child. There are 9 people in crafts class today. How many colored sheets of paper will the crafts teacher need today?

10 points

⟨Ans.⟩ _____

5 We had 6 pieces of tape that were each 5 inches long. Just for fun, we connected them end-to-end. How long was our new piece of tape?

10 points

| 5 in. | 5 in. | 5 in. | 5 in. | 5 in. | 5 in. |

⟨Ans.⟩ _____

6 Nathalie went to the supply closet where her mother worked and found 3 boxes of pencils. Each box was almost empty and had only 5 pencils per box. How many pencils were there altogether?

10 points

Number of pencils per box		Number of boxes		Total pencils
☐	×	☐	=	☐

⟨**Ans.**⟩ _____

7 Sean helped his mother take the groceries inside. He picked up 3 boxes that each held 5 apples. How many apples did he pick up in all?

10 points

Number of apples per box		Number of boxes		Total apples
☐	×	☐	=	☐

⟨**Ans.**⟩ _____

8 The milkman was making his rounds. He had 5 boxes which each held 4 bottles of milk. How many bottles of milk did he have?

10 points

Number of bottles per box		Number of boxes		Total bottles
☐	×	☐	=	☐

⟨**Ans.**⟩ _____

9 At the beginning of the test, the teacher gives 5 pencils to each child. If there are 8 people in the class, how many pencils will the teacher need?

10 points

⟨**Ans.**⟩ _____

10 The length of the flowerbed is 4 meters. The width is three times the length. How wide is the flowerbed?

10 points

4 m

------- 3 times the length -------

⟨**Ans.**⟩ _____

Are you getting the hang of it? Good!

11 **Multiplication**

Level ★ ★

Date   /   /

Name

Score

/100

1 Sheila is going home for the holidays, so she bought 4 boxes of candy. Each box contains 6 candies. How many candies did she buy?

10 points

⟨Ans.⟩ _____

2 You want to cut enough tape to fasten 3 ribbons to the present you are wrapping. Each ribbon needs 6 centimeters of tape. How much tape do you cut in all?

10 points

⟨Ans.⟩ _____

3 Grandfather always gives his grandchildren money when they come to see him. This winter, he gave 7 coins to each grandchild. If he has 5 grandchildren, how many coins did he need?

10 points

⟨Ans.⟩ _____

4 In the school closet, there are boxes with pens. If there are 7 pens in each box, and there are 4 boxes, how many pens are there in the closet?

10 points

⟨Ans.⟩ _____

5 One week is 7 days. How many days are in three weeks?

10 points

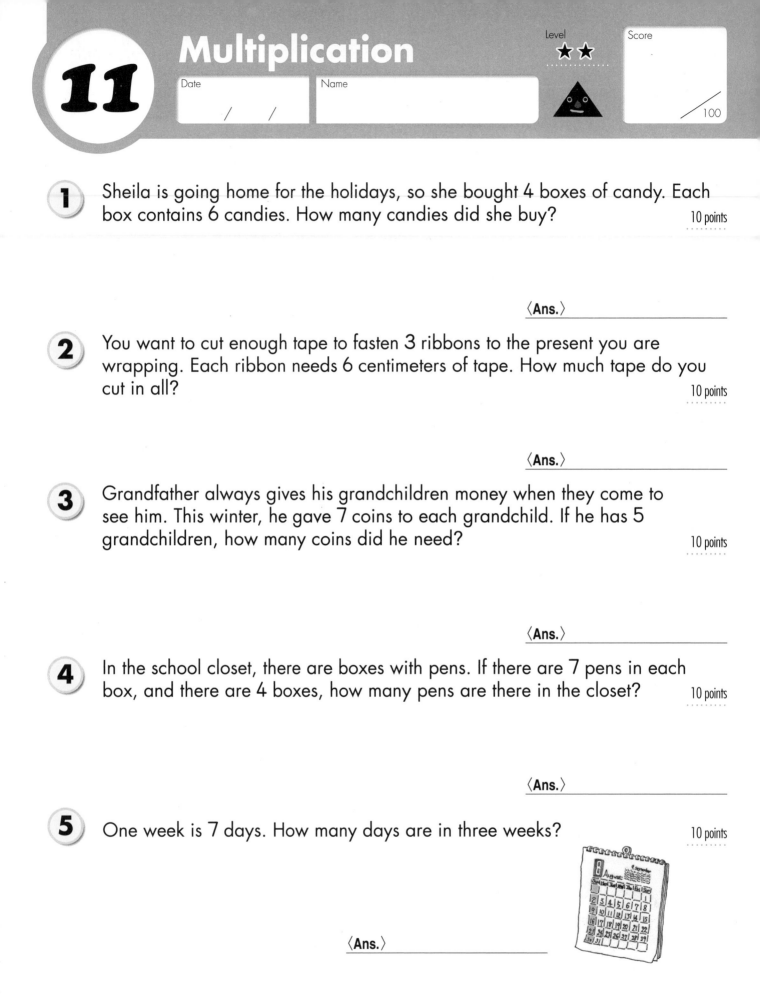

⟨Ans.⟩ _____

6 When the ferry got stuck, everyone had to get in the lifeboats. 6 people fit in each lifeboat, and there were 5 lifeboats. How many people could get in the lifeboats at one time?

10 points

⟨**Ans.**⟩ _____

7 Gina really likes the paddleboats so she made us all go. There were 5 boats and 6 people could get in each boat. How many people could go on the paddleboats?

10 points

□ × □ = □

⟨**Ans.**⟩ _____

8 There are 6 children in your carpool, and you want to give them each 7 pieces of candy. How many pieces of candy will you need?

10 points

⟨**Ans.**⟩ _____

9 The cafeteria has 9 boxes filled with bottles of juice. If there are 6 bottles of juice per box, how many bottles are there in all?

10 points

⟨**Ans.**⟩ _____

10 Ben was in the back playing with his father's bricks. He piled up 7 bricks, one on top of the other. If each brick was 8 centimeters thick, how tall was his pile?

10 points

⟨**Ans.**⟩ _____

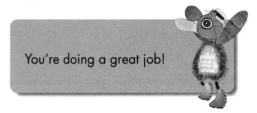

You're doing a great job!

1 On Valentine's Day, you gave 5 people in your class 8 pieces of chocolate each. How many chocolates did you give away?

10 points

⟨Ans.⟩

2 Each baseball team is made up of 9 players. If there are 7 teams waiting to play, how many players are there altogether?

10 points

⟨Ans.⟩

3 I bought flowers for all 3 of my teachers at the end of the year. If each bunch of flowers included 8 flowers, how many flowers did I give away?

10 points

⟨Ans.⟩

4 Peggy measured the hallway with an 8-meter tape. She found that the length of the hallway was 6 times the length of her tape. How long was the hallway?

10 points

⟨Ans.⟩

5 You used 4 sticks to make a square. If the sticks were 9 centimeters long, how long is it around the square?

10 points

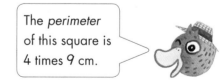

The *perimeter* of this square is 4 times 9 cm.

⟨Ans.⟩

6 Lora was working in the stock room, and she piled up 7 boxes. Each box was 8 centimeters high. How high was the stack of boxes?

10 points

⟨Ans.⟩ _____

7 You bought 5 sheets of drawing paper. One sheet cost 9¢. How much should you pay for all 5 sheets?

10 points

☐ × ☐ = ☐

⟨Ans.⟩ _____

8 In Julio's class, there are 4 groups, and each group has 8 students. How many students are in his class?

10 points

⟨Ans.⟩ _____

9 There are 9 couches in the student lounge, and 8 people can sit on each couch. How many people can sit down in the student lounge in all?

10 points

⟨Ans.⟩ _____

10 Gordon had a big birthday party. He had 3 cakes, and he cut each cake into 9 pieces. How many pieces of cake did Gordon have?

10 points

⟨Ans.⟩ _____

Getting tougher right? Stick with it!

Multiplication

1 In Carol's class, they split into groups of 6 for an exercise. If there were 6 people in each group, how many people are in Carol's class? 10 points

⟨**Ans.**⟩ _____

2 Ted's family was gathering around the fire. He put 8 chestnuts on a dish for each of them. If there are 4 people in Ted's family, how many chestnuts were there in all? 10 points

☐ × ☐ = ☐

⟨**Ans.**⟩ _____

3 You brought 6 pencils for each person in your study group. If there are 7 children in your study group, how many pencils will you need? 10 points

⟨**Ans.**⟩ _____

4 You want to buy 8 stamps and each stamp is 5¢. How much money do you need to bring? 10 points

⟨**Ans.**⟩ _____

5 The local store has 3 boxes of soap, each containing 6 pieces of soap. How many pieces of soap does the local store have? 10 points

⟨**Ans.**⟩ _____

6 You are playing tangrams by yourself. You want to arrange your triangles as in the figure below. If you want to make 5 more identical figures, how many triangles will you need?

10 points

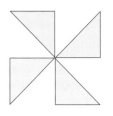

⟨**Ans.**⟩ _____

7 Vinny was having fun playing with tape. He connected 8 pieces of tape that were 7 centimeters long. How long was his new piece of tape?

10 points

⟨**Ans.**⟩ _____

8 We were playing basketball in the park, and we made teams with 5 people in each team. If there were 9 teams, how many people were playing basketball?

10 points

⟨**Ans.**⟩ _____

9 Dara put up flags for the cross-country race. She put up 8 flags. If there was a distance of 3 meters between every flag, how far was it from the first flag to the eighth flag?

10 points

⟨**Ans.**⟩ _____

10 Matt's grandmother is sick. She got 9 bunches of flowers, and each bunch had 7 flowers in it. How many flowers did Matt's grandmother get?

10 points

⟨**Ans.**⟩ _____

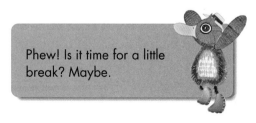

Phew! Is it time for a little break? Maybe.

Multiplication

14

Date   /   /

Name

Score   /100

1 A packet of colored paper includes 30 sheets. You bought 4 packets. How many sheets did you buy?

10 points

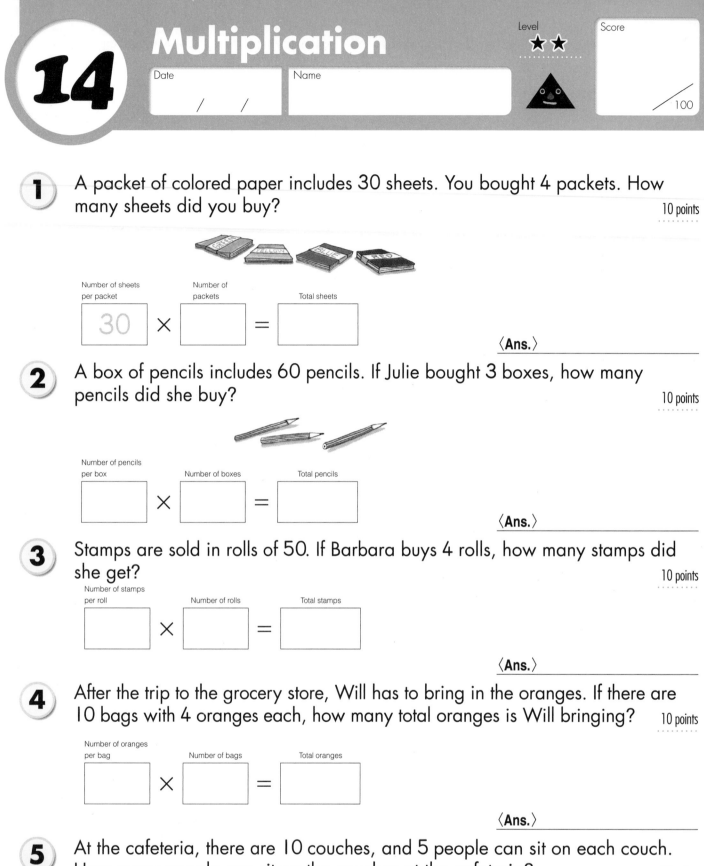

Number of sheets per packet | Number of packets | Total sheets

$\boxed{30}$ × $\boxed{}$ = $\boxed{}$

⟨**Ans.**⟩ _____

2 A box of pencils includes 60 pencils. If Julie bought 3 boxes, how many pencils did she buy?

10 points

Number of pencils per box | Number of boxes | Total pencils

$\boxed{}$ × $\boxed{}$ = $\boxed{}$

⟨**Ans.**⟩ _____

3 Stamps are sold in rolls of 50. If Barbara buys 4 rolls, how many stamps did she get?

10 points

Number of stamps per roll | Number of rolls | Total stamps

$\boxed{}$ × $\boxed{}$ = $\boxed{}$

⟨**Ans.**⟩ _____

4 After the trip to the grocery store, Will has to bring in the oranges. If there are 10 bags with 4 oranges each, how many total oranges is Will bringing?

10 points

Number of oranges per bag | Number of bags | Total oranges

$\boxed{}$ × $\boxed{}$ = $\boxed{}$

⟨**Ans.**⟩ _____

5 At the cafeteria, there are 10 couches, and 5 people can sit on each couch. How many people can sit on the couches at the cafeteria?

10 points

⟨**Ans.**⟩ _____

6 In our party today at school, everyone is supposed to get 8 candies. If we have 30 people in the class, how many candies do we need?

10 points

Number of candies per person
[8] × Number of people [ ] = Total candies [ ]

⟨Ans.⟩ _____

7 Robin loves being a florist. Today, she is tying bunches of flowers together, with 6 flowers in each bunch. If she is making 40 bunches, how many flowers should she start with?

10 points

Number of flowers per bunch
[ ] × Number of bunches [ ] = Total flowers [ ]

⟨Ans.⟩ _____

8 The gardener gave each child 5 seeds to plant. If there are 32 children, how many seeds did the gardener give away?

10 points

⟨Ans.⟩ _____

9 Each pack of crayons has 20 crayons in it. Your class has 18 packs. How many crayons does your class have?

10 points

⟨Ans.⟩ _____

10 There are 36 people in Tim's class. If he wants to give them each 3 sheets of paper, how many sheets will he need?

10 points

⟨Ans.⟩ _____

Okay, looks like you've got it. Now let's make it a little harder!

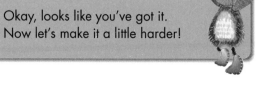

Multiplication

15

Level ★★

Score
/100

Date   /   /

Name

1 There are packets with 4 candies inside each packet. If Carla gives 2 packets each to 5 people, how many candies is she giving away?   12 points

Number of candies per packet   Number of packets   Number of people   Total candies

$$\boxed{4} \times \boxed{2} \times \boxed{5} = \boxed{}$$

⟨Ans.⟩ _____

2 Mother was packing the groceries and put 5 oranges in each bag. If she gave 3 of these bags each to 3 people, how many oranges did she buy?   12 points

⟨Ans.⟩ _____

3 You are at the corner store to buy some snacks. There is a snack box that contains 5 bags of pretzels. Each bag of pretzels has 70 pretzels inside. If you buy 2 boxes, how many pretzels will you get?   12 points

Number of pretzels per bag   Number of bags   Number of boxes   Total pretzels

$$\boxed{} \times \boxed{} \times \boxed{} = \boxed{}$$

⟨Ans.⟩ _____

4 Jerry is looking at an art set that contains 2 sheets of stickers. Each sheet has 45 stickers on it. If he buys 3 art sets, how many stickers will he get?   12 points

⟨Ans.⟩ _____

5 At the supermarket, there are 3 kinds of boxes—small, medium and large. The small box contains 3 cakes. The medium box holds 2 times as many cakes as small box, and the large box holds 4 times as many cakes as the middle box. How many cakes are in the big box?

13 points

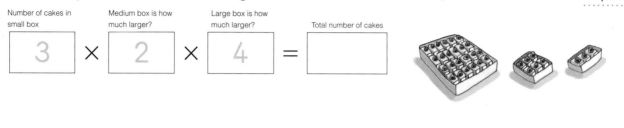

Number of cakes in small box		Medium box is how much larger?		Large box is how much larger?		Total number of cakes
3	×	2	×	4	=	

⟨Ans.⟩ _____

6 Kim, Mary and Julia were knitting together. Kim knitted 8 centimeters of string. Mary knitted 2 times as much as Kim and Julia knitted 3 times as much as Mary. How much string did Julia knit?

13 points

⟨Ans.⟩ _____

7 James is sick and has to take 2 different pills. He has to take them both 3 times each day. If he has to keep taking the medicine for 5 days, how many pills will he take?

13 points

⟨Ans.⟩ _____

8 Benji is also sick, but he has to take 3 different pills. He has to take them 3 times each day, for 4 days. How many pills will he take?

13 points

⟨Ans.⟩ _____

That was tough, but you stuck with it! Nice!

Division

16

Date / /

Name

Level ★★

Score /100

1 If 2 people get to share 6 candies equally, how many candies will each person get?

10 points

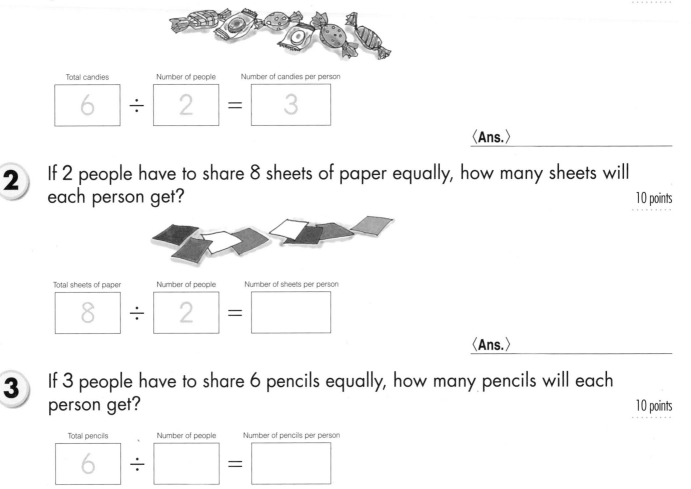

Total candies		Number of people		Number of candies per person
6	÷	2	=	3

〈Ans.〉 _____

2 If 2 people have to share 8 sheets of paper equally, how many sheets will each person get?

10 points

Total sheets of paper		Number of people		Number of sheets per person
8	÷	2	=	

〈Ans.〉 _____

3 If 3 people have to share 6 pencils equally, how many pencils will each person get?

10 points

Total pencils		Number of people		Number of pencils per person
6	÷		=	

〈Ans.〉 _____

4 On Halloween night, Mrs. Kwan had 12 candies left. 3 children came to the door, and she shared the candies equally among their bags. How many candies did each child get?

10 points

Total candies		Number of children		Number of candies per child
	÷		=	

〈Ans.〉 _____

5 We have 15 pencils for my group today. If the 3 of us share them equally, how many pencils will I get?

10 points

Total pencils		Number of people		Number of pencils per person
15	÷		=	

⟨Ans.⟩ _____

6 Yesterday, we had 15 pencils for my group, too, but we had 5 people in the group. If we shared the pencils equally yesterday, how many pencils did I get then?

10 points

Total pencils		Number of people		Number of pencils per person
	÷		=	

⟨Ans.⟩ _____

7 Gina is putting together gift bags for Valentine's Day. She has 18 chocolates and she divided them into the 3 bags equally. How many chocolates will she put in each bag?

10 points

⟨Ans.⟩ _____

8 In your group for crafts class, you have 4 people. If you are supposed to share 24 sheets of paper equally, how many sheets of paper will each of you get?

10 points

⟨Ans.⟩ _____

9 If you divide 25 centimeters of ribbon into 5 equal parts, how long would each of those parts be?

10 points

⟨Ans.⟩ _____

10 Mother is helping us pack for the camping trip. We have 28 oranges and 7 backpacks. How many oranges should we put in each backpack?

10 points

⟨Ans.⟩ _____

Getting a handle on these division word problems? Good!

33

Division

17

Date / /

Name

Level ★★

Score

/100

---

1 You have 6 candies. How many people will get candies if you give them 2 candies each?

10 points

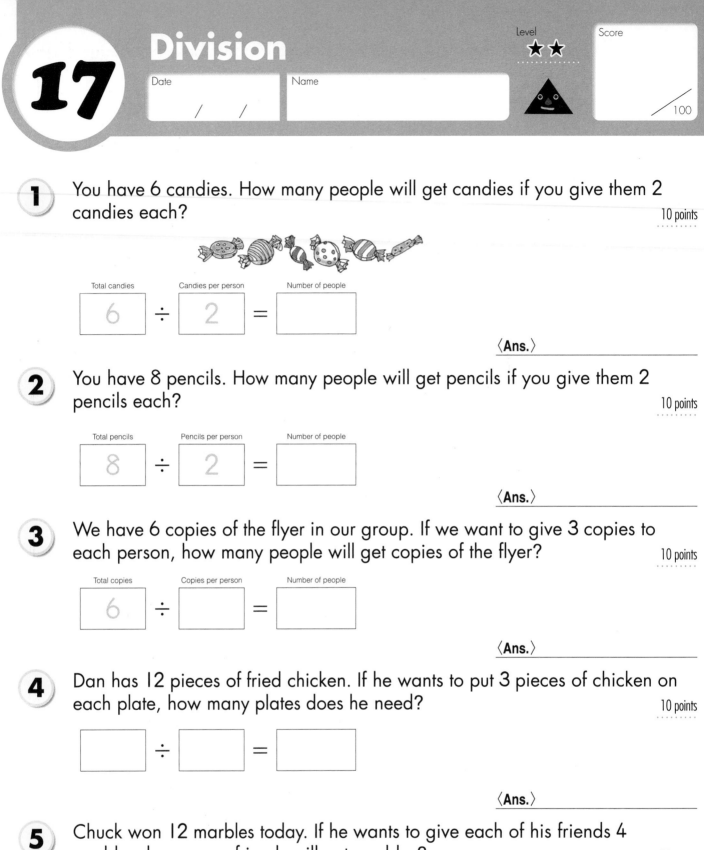

Total candies
6

÷

Candies per person
2

=

Number of people

⟨Ans.⟩ _____

2 You have 8 pencils. How many people will get pencils if you give them 2 pencils each?

10 points

Total pencils
8

÷

Pencils per person
2

=

Number of people

⟨Ans.⟩ _____

3 We have 6 copies of the flyer in our group. If we want to give 3 copies to each person, how many people will get copies of the flyer?

10 points

Total copies
6

÷

Copies per person

=

Number of people

⟨Ans.⟩ _____

4 Dan has 12 pieces of fried chicken. If he wants to put 3 pieces of chicken on each plate, how many plates does he need?

10 points

÷

=

⟨Ans.⟩ _____

5 Chuck won 12 marbles today. If he wants to give each of his friends 4 marbles, how many friends will get marbles?

10 points

⟨Ans.⟩ _____

6 Abby has 15 sodas to give away to her friends. If she gives away 3 sodas per person, how many friends can she give soda to?

10 points

Total sodas		Sodas per person		Number of friends
15	÷		=	

⟨Ans.⟩ _____

7 Abby still has 15 sodas to give away, but now she wants to give her friends 5 sodas each. How many friends get soda now?

10 points

Total sodas		Sodas per person		Number of friends
	÷		=	

⟨Ans.⟩ _____

8 Your group in crafts class has 18 sheets of craft paper. If you want to give each person 3 sheets, how many people will get craft paper?

10 points

⟨Ans.⟩ _____

9 Mrs. Williams is dividing up 24 flowers into different vases. If she puts 6 flowers into each vase, how many vases will she need?

10 points

⟨Ans.⟩ _____

10 Denise has a 32-inch ribbon but needs to divide it up into 8-inch pieces. How many 8-inch pieces can she make?

10 points

⟨Ans.⟩ _____

Good job!!

Division

Level ★★

Score /100

1 We have 21 pencils for class today. If we wanted to give 3 people all of the pencils in equal amounts, how many pencils would each person get?   10 points

☐ ÷ ☐ = ☐

⟨**Ans.**⟩ _____

2 We still have 21 pencils for class today. If we give 3 pencils to each person, how many people would get pencils?   10 points

☐ ÷ ☐ = ☐

⟨**Ans.**⟩ _____

3 Father is trying to save 30 liters of oil in 5 different cans. If he divides the oil equally, how much oil will there be in one can?   10 points

⟨**Ans.**⟩ _____

4 One set of 8 bamboo sticks at the local flower shop is 40¢. How much is 1 bamboo stick?   10 points

⟨**Ans.**⟩ _____

5 Mrs. Kemp has 72 pieces of chicken, but she wants to save some in the freezer. If she divides them into packs of 8 pieces each, how many packs of chicken will she make?   10 points

⟨**Ans.**⟩ _____

6 Your neighbor is having a yard sale. They are selling toy soldiers for 8¢ each. If you have 40¢, how many soldiers can you buy?

10 points

⟨**Ans.**⟩ _____

7 Your father got 30 kilograms of potatoes from his uncle, who is a farmer. If he gives 5 kilograms of potatoes to each neighbor on your street, how many neighbors will get potatoes?

10 points

⟨**Ans.**⟩ _____

8 Amelia has 20 pages of homework this week. If she does 5 pages per day, how many days will it take her to finish her homework?

10 points

⟨**Ans.**⟩ _____

9 There are 18 bottles of juice for the class party today. The teacher asked Steve and Brian to carry them into class. If they divide up the bottles evenly, how many bottles will they carry each?

10 points

⟨**Ans.**⟩ _____

10 Flo has 6 of her textbooks in a pile. Each book is the same thickness, and the height of the pile is 5 centimeters 4 millimeters. How thick is each book?

10 points

> 1 cm = 10 mm
> 5 cm 4 mm = 54 mm

⟨**Ans.**⟩ _____

Wow! You're doing division with measurements. Good job!

1 Today is pirate day at school, and Cindy's team has 20 coins. Her group has 3 people, so they divide the coins equally and everyone gets 6 coins. How many coins remain? 10 points

$$20 \div 3 = 6 \text{ R } 2 \quad \text{Remainder}$$

⟨Ans.⟩ ☐ coins remain

2 In crafts class today, our group got 30 colored sheets of paper. Each of the 4 people in the group got 7 sheets. How many sheets of paper remain? 10 points

$$30 \div 4 = 7 \text{ R } \boxed{}$$

⟨Ans.⟩ ☐ sheets remain

3 Adam's mother is making lunch bags. She divides 27 oranges into 6 bags evenly. How many oranges are in each bag, and how many remain? 10 points

$$27 \div \boxed{} = \boxed{} \text{ R } \boxed{}$$

⟨Ans.⟩ ☐ oranges each bag, ☐ oranges remain.

4 The Miller family got 40 pieces of chocolate for Christmas. If there are 7 children in the Miller family, how many pieces of chocolate does each child get, and how many are left over? 10 points

⟨Ans.⟩ _____

5 In Mr. Shibata's pond, there are 27 fish. He's cleaning the pond today and divides the fish equally into 6 buckets. How many fish are in each bucket, and how many are left over? 10 points

⟨Ans.⟩ _____

6 There are 20 slices of pizza at Randy's party. If 6 people get 3 slices each, how many slices remain?

10 points

$$\boxed{20} \div \boxed{3} = \boxed{} \text{ R } \boxed{}$$

⟨**Ans.**⟩ $\boxed{}$ slices remain

7 At the dog pound, Cristina has 30 treats to give to her dogs. If she gives 4 each to her 7 dogs, how many treats remain?

10 points

$$\boxed{30} \div \boxed{4} = \boxed{} \text{ R } \boxed{}$$

⟨**Ans.**⟩ $\boxed{}$ treats remain

8 Your mother is making lunch bags for you. She has 36 rice cakes and puts 8 in each bag. How many bags will she make, and how many cakes will remain?

10 points

$$\boxed{} \div \boxed{} = \boxed{} \text{ R } \boxed{}$$

⟨**Ans.**⟩ $\boxed{}$ bags, $\boxed{}$ cakes remain.

9 We are packing presents, and we have 75 inches of ribbon. If we use 8 inches of ribbon on each present, how many presents can we wrap? How much ribbon will be left over?

10 points

⟨**Ans.**⟩ _____

10 Susan sells filled picnic baskets at her store. She has 65 apples. If she puts 7 apples into each picnic basket, how many baskets can she make? How many apples will she have remaining?

10 points

⟨**Ans.**⟩ _____

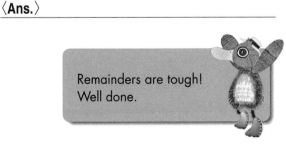

Remainders are tough! Well done.

1 Tim quits playing marbles and gives us all of his marbles. He has 45 marbles, and he gives 6 marbles to each person. How many people got marbles, and how many marbles are left over?

10 points

⟨Ans.⟩ ☐ people, ☐ marbles remain.

2 Lisa is working at the flower shop this weekend. She has 60 carnations, and is supposed to put 9 in each vase. How many vases can she make? How many carnations will she have left?

10 points

⟨Ans.⟩ ☐ vases, ☐ carnations remain.

3 In math class today, we got 38 sheets of graph paper and had to give equal amounts to 8 different groups. How many sheets did each group get, and how many sheets were left over?

10 points

⟨Ans.⟩ _____

4 In crafts class, we are folding paper cranes in groups of 7 today. Ted's group got 60 colored sheets. If everyone in his group folded the same amount of cranes, how many cranes did they fold, and how many sheets of paper were left?

10 points

⟨Ans.⟩ _____

5 The pet store has 38 turtles. If they can only put 6 turtles in a tank together, how many tanks will they need, and how many turtles will remain?

10 points

⟨Ans.⟩ _____

6 Amy is using thumbtacks to put up her favorite pictures. She has 25 thumbtacks, and hanging each picture takes 4 thumbtacks. How many pictures can she hang?

10 points

〈Ans.〉

7 I have a bookshelf exactly like the one shown below. If my books are 6 centimeters thick, how many books can I fit on my bookshelf?

10 points

50cm

〈Ans.〉

8 The boats at Ron's summer camp fit 4 people each. If there are 26 people at the pond today, how many boats do they need?

10 points

〈Ans.〉

9 You can make 8 greeting cards from 1 sheet of paper. If you want to make 30 greeting cards, how many sheets of paper will you need?

10 points

〈Ans.〉

10 Sonia can fit 6 tennis balls in each of her containers. If she has 50 tennis balls, how many containers does she need?

10 points

〈Ans.〉

Way to go! Now let's try something different.

1 Victor is a parking lot attendant. He watched 6 cars leave, and then 7 more cars left. If there are 15 cars now, how many cars were there at first?

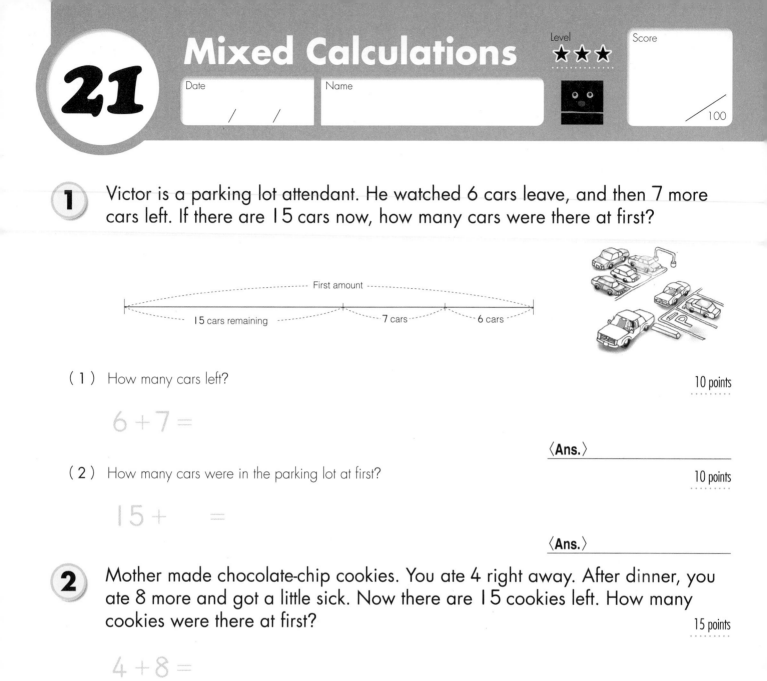

First amount

15 cars remaining   7 cars   6 cars

(1) How many cars left?   10 points

$6 + 7 =$

⟨Ans.⟩ _____

(2) How many cars were in the parking lot at first?   10 points

$15 + \quad =$

⟨Ans.⟩ _____

2 Mother made chocolate-chip cookies. You ate 4 right away. After dinner, you ate 8 more and got a little sick. Now there are 15 cookies left. How many cookies were there at first?   15 points

$4 + 8 =$

⟨Ans.⟩ _____

3 A flock of sparrows is hanging out on Farmer Hampton's fence. When his dog ran after them, 5 flew away. When Farmer Hampton went to the fence to retrieve his dog, 7 more flew away. There are still 12 sparrows on the fence. How many sparrows were on the fence at first?   15 points

⟨Ans.⟩ _____

4 A gaggle of geese is swimming around the pond in the zoo. It was getting cold, so 14 geese got out, and then 3 more got out of the pond. 9 geese are still in the pond. How many geese were there in all?

10 points

⟨**Ans.**⟩

5 Mother made apple pie for the new people in the neighborhood. You gave 8 slices to one of the neighbors and 11 to another. If you have 18 slices left, how many were there at first?

10 points

⟨**Ans.**⟩

6 We were making paper dolls in craft class, and we used 15 sheets of colored paper for the bodies. Then we used 10 sheets for the arms and legs. If we have 35 sheets of colored paper left, how many sheets were there at first?

10 points

⟨**Ans.**⟩

7 Jim is on the bus to his grandparents' house. 11 people got off the bus at the library, and 13 got off at the train station. Now there are 12 people on Jim's bus. How many people were there at the beginning?

10 points

⟨**Ans.**⟩

8 Maria is using some ribbon to wrap presents. Yesterday, she used 50 inches of ribbon, and today, she used 40 inches. If she still has 130 inches left, how much ribbon did she have at first?

10 points

⟨**Ans.**⟩

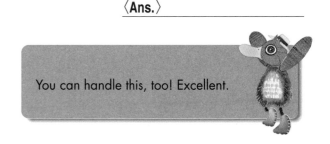

You can handle this, too! Excellent.

1 Greg is shopping for a picnic. He bought blueberries and raspberries. There are 60 blueberries and 80 raspberries in his basket. He added blackberries at the last minute, and he had 210 berries altogether. How many blackberries did he buy?

10 points per question

(1) How many blueberries and raspberries did he have altogether?

$$60 + 80 =$$

⟨**Ans.**⟩ _____

(2) How many blackberries did he buy?

$$210 - \quad =$$

⟨**Ans.**⟩ _____

2 The students were cleaning up after art class. They put 70 crayons and 40 markers in the art supply closet. The teacher put away the paintbrushes, and all 230 art supplies were put away. How many paintbrushes did the class use?

10 points per question

(1) How many crayons and markers were there altogether?

⟨**Ans.**⟩ _____

(2) How many paintbrushes did the class us?

⟨**Ans.**⟩ _____

3 In crafts class today, we got 36 red sheets of paper and 25 yellow sheets of paper. The teacher left and came back with some blue sheets of paper, and now the class has 82 sheets of paper. How many blue sheets of paper did we get?

10 points

⟨**Ans.**⟩ _____

4 Libby's father was planning her birthday party. He bought 80 stickers and 75 bouncing balls. He also bought candies, and altogether he had 270 party favors. How many candies did he buy?

10 points

⟨**Ans.**⟩ _____

5 We were playing tag in the park with 14 girls and 13 boys. Then some girls joined in, and we had 33 people playing altogether. How many girls joined in?

10 points

⟨**Ans.**⟩ _____

6 Nancy got 18 stickers from her sister and 23 stickers from her brother. If her mother gave her some stickers too, and Nancy has 60 stickers in all, how many stickers did her mother give her?

15 points

⟨**Ans.**⟩ _____

7 Tom and Sam were making paper airplanes today after class. Tom used 16 sheets of paper and Sam used 15. Then Ted joined in and made some. In all, they used 48 sheets of paper. How many sheets did Ted use?

15 points

⟨**Ans.**⟩ _____

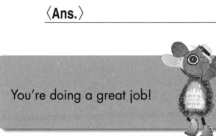

You're doing a great job!

1 You brought some stickers to class for Valentine's Day. You gave 4 people 5 stickers each, and you had 30 stickers left. How many did you have at first?

10 points per question

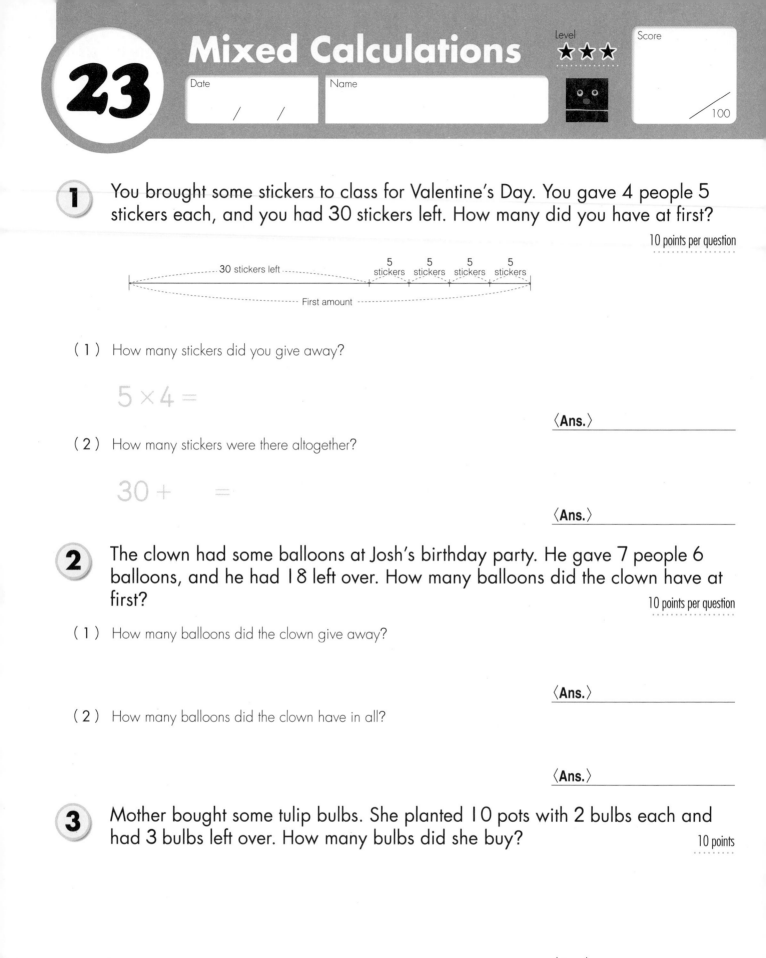

30 stickers left

5 stickers 5 stickers 5 stickers 5 stickers

First amount

(1) How many stickers did you give away?

$5 \times 4 =$

⟨**Ans.**⟩ _____

(2) How many stickers were there altogether?

$30 + \quad =$

⟨**Ans.**⟩ _____

2 The clown had some balloons at Josh's birthday party. He gave 7 people 6 balloons, and he had 18 left over. How many balloons did the clown have at first?

10 points per question

(1) How many balloons did the clown give away?

⟨**Ans.**⟩ _____

(2) How many balloons did the clown have in all?

⟨**Ans.**⟩ _____

3 Mother bought some tulip bulbs. She planted 10 pots with 2 bulbs each and had 3 bulbs left over. How many bulbs did she buy?

10 points

⟨**Ans.**⟩ _____

4 Chuck works for the amusement park and gave away some coupons. He gave 8 people 4 coupons each and still had 21 coupons left. How many coupons did Chuck have originally?

10 points

〈**Ans.**〉 _____

5 Lee's mother bought some carrots to make carrot soup. She used 7 bushels of carrots that each had 20 carrots. She had 60 carrots left over. How many carrots did she buy?

10 points

〈**Ans.**〉 _____

6 Mr. Myers is preparing for the winter, so he went and bought 6 boxes of logs for the fire. Each box has 25 logs in them. After he checked the boxes when he got home, he found that 8 logs were missing. How many logs does he have?

10 points

〈**Ans.**〉 _____

7 At the bakery, you bought 6 buns that cost 76¢ each. If you paid with $5, how much money did you get back?

10 points

$5 = 500¢

〈**Ans.**〉 _____

8 Rachel bought 3 hair clips that cost 65¢ each. She paid with $2. How much change did she get?

10 points

〈**Ans.**〉 _____

Just put one foot in front of the other. Well done!

47

1 For our big holiday dinner, we put 3 small desserts on each dish. We made 27 desserts, and there are 5 dishes left. How many dishes do we have in all?

10 points per question

(1) How many dishes were used for the desserts?

$27 \div 3 =$

⟨Ans.⟩ _____

(2) How many dishes are there altogether?

$5 + \qquad =$

⟨Ans.⟩ _____

2 Michele planted 2 seeds in each pot. She planted 16 seeds, and there are still 4 pots left. How many pots are there in all?

10 points per question

(1) How many pots have seeds planted in them?

☐ ÷ ☐ = ☐

⟨Ans.⟩ _____

(2) How many pots are there altogether?

☐ + ☐ = ☐

⟨Ans.⟩ _____

3 Dan's father is an orange farmer, and he brought 32 oranges to class today. If each child got 4 oranges, and 5 children didn't want oranges, how many children are in the class?

10 points

⟨Ans.⟩ _____

4 Farmer Daniels got 72 eggs from the hens today. He put 8 in each box, and sold 2 boxes right away. How many boxes of eggs does he have left? 10 points

$$\boxed{} \div \boxed{} = \boxed{}$$

$$\boxed{} - \boxed{2} = \boxed{}$$

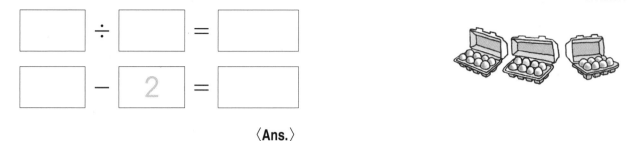

⟨Ans.⟩ _____

5 Farmer Daniels also harvested his peaches today. He got 54 peaches and put 6 in each basket. If he started with 12 baskets, how many empty baskets does he have left? 10 points

⟨Ans.⟩ _____

6 Patty had 25 stickers. Today, she got 18 more, but her mother told her to split those evenly with her sister. How many stickers does Patty have now? 15 points

⟨Ans.⟩ _____

7 Mike had 36 pennies yesterday. Today, his father gave him 21 pennies, but told him to split those with his brother and sister evenly. How many pennies does Mike have now? 15 points

⟨Ans.⟩ _____

This is tough! Good job!

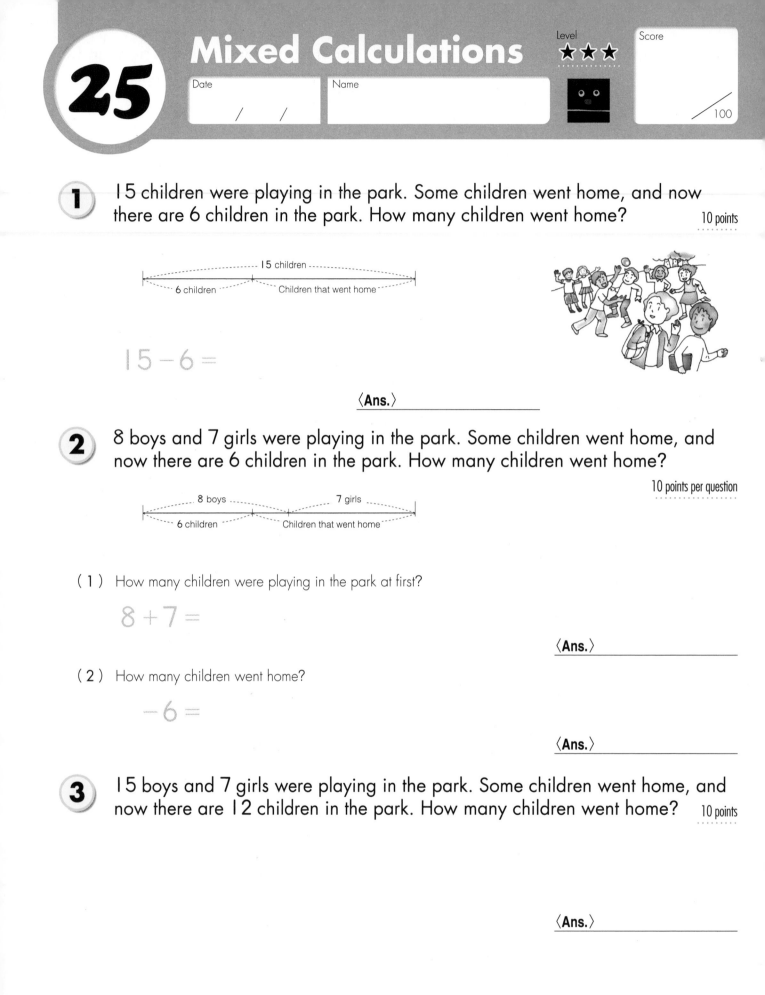

Mixed Calculations

Level ★★★

Date   /   /

Name

Score
/100

1 15 children were playing in the park. Some children went home, and now there are 6 children in the park. How many children went home?

10 points

15 children
6 children — Children that went home

$15 - 6 =$

⟨Ans.⟩ _____

2 8 boys and 7 girls were playing in the park. Some children went home, and now there are 6 children in the park. How many children went home?

10 points per question

8 boys — 7 girls
6 children — Children that went home

(1) How many children were playing in the park at first?

$8 + 7 =$

⟨Ans.⟩ _____

(2) How many children went home?

$- 6 =$

⟨Ans.⟩ _____

3 15 boys and 7 girls were playing in the park. Some children went home, and now there are 12 children in the park. How many children went home?

10 points

⟨Ans.⟩ _____

4 8 boys and 12 girls were playing basketball. It started getting dark, so some of them left. Now, there are only 7 children playing basketball. How many children went home?

15 points

⟨Ans.⟩ _____

5 Mother had 13 red flowers and 9 white flowers in the yard. She cut some and put them in a vase above the fireplace. Now she only has 8 flowers left. How many flowers did she cut?

15 points

⟨Ans.⟩ _____

6 Lucy had 22 yellow candies and 15 blue candies. She was bored all day and ate some of her candies, and now she has 19 candies in all. How many candies did she eat today?

15 points

⟨Ans.⟩ _____

7 17 ducks and 7 swans swim around the pond at the zoo. If some of them left the pond, and now there are 11 of them in the pond, how many ducks and swans got out of the pond?

15 points

⟨Ans.⟩ _____

I never knew candy could be so difficult. Way to go!

1 Grandmother has 6 grandchildren. She picked 24 wild strawberries today and gave each of her grandchildren an equal amount. How many strawberries did each grandchild get?

10 points

6 children

24 strawberries

$24 \div 6 =$ 〈Ans.〉 _____

2 In my family, there are 3 brothers and 2 sisters. My mother had 25 oranges recently, and she divided them equally among us. How many oranges did each of us get?

10 points per question

3 brothers   2 sisters

25 oranges

(1) How many children are there in my family?

$3 + 2 =$ 〈Ans.〉 _____

(2) How many oranges did each child get?

$25 \div \quad =$ 〈Ans.〉 _____

3 There are 3 boys and 4 girls in Peter's group for this project. His group got 21 sheets of paper and divided them equally. How many sheets of paper did each get?

10 points

〈Ans.〉 _____

4 Allison bought 30 oranges. As she was checking out, the grocery store was running out of bags. They only had 4 paper bags and 2 plastic bags. If they divided them up equally, how many oranges did they put in each bag?

10 points

〈Ans.〉 _____

5 You brought some candy to school, and you gave 2 candies to each of your 4 friends. If the candies cost you 40¢ altogether, how much did each candy cost you?

10 points per question

```
    2           2           2           2
```
40¢

(1) How many candies did you give away?

$2 \times 4 =$

〈Ans.〉 _____

(2) How much did each candy cost?

$40 \div \quad =$

〈Ans.〉 _____

6 Laura gave 3 pieces of gum each to her 2 best friends. If she bought those pieces of gum for 42¢ in all, how much did each piece of gum cost?

10 points

〈Ans.〉 _____

7 At the baseball game, Tina, Ted and Tom each got 2 bags. Then, their father divided 48 peanuts equally among the bags. How many peanuts were in each bag?

10 points

〈Ans.〉 _____

8 Holly gave 3 hair clips to each of her 3 best friends. If she bought all of the hair clips for 45¢ total, how much did each hair clip cost?

10 points

〈Ans.〉 _____

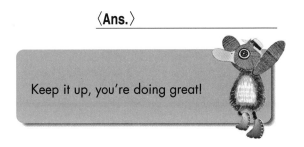
Keep it up, you're doing great!

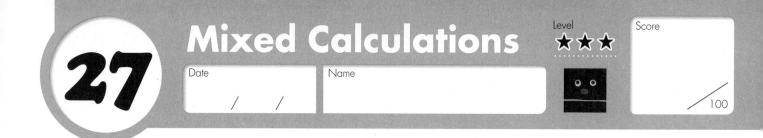

Mixed Calculations

27

Level ★★★

Date   /   /

Name

Score
/100

1 We bought oranges today and put 6 oranges each in our 4 bags. When we get to the party, we are going to divide these oranges among 8 people equally. How many oranges will each person get at the party?

10 points

$6 \times 4 = 24$

$24 \div 8 =$

⟨Ans.⟩ _____

2 A group of 6 children bought 3 dozen pencils together in order to save money. If they divide the 3 dozen equally, how many pencils will each person in the group get?

10 points

⟨Ans.⟩ _____

3 It's very crowded at the hopscotch area today. There are 3 lines with 8 girls in each line. They decided to draw another hopscotch board, and now there are 4 equal lines. How many girls are in each line now?

10 points

⟨Ans.⟩ _____

4 Our class project is 2 strings of paper cranes for the local hospital. We have 8 children making paper cranes. If we need 36 cranes for each string, how many cranes does each child have to make?

10 points

⟨Ans.⟩ _____

5 Carmen likes working at the florist. Today, she made 4 bunches of flowers with 6 flowers each. Her boss said she needed 8 flowers in each bunch, though. How many bunches can she make using the same amount of flowers?

10 points

⟨Ans.⟩ _____

6 Thomas bought 3 toy soldiers for 27¢. How much would it cost him to buy 5 toy soldiers?

10 points

$27 \div 3 =$

⟨Ans.⟩ _____

7 Gina was weighing her mother's rings. She found 3 rings that were the same weight, and they weighed 24 grams in all. If she found 7 rings that weighed this same weight, how much would the weigh in all?

10 points

⟨Ans.⟩ _____

8 Today, Lana is selling flower bunches at the park. She divided 45 flowers into equal bunches of 5 flowers each. If she sells each bunch for $3 each, how much money will she earn?

10 points

⟨Ans.⟩ _____

9 Eric had 18 chocolate bars, but he ate half of them. If he eats the other half over the next 3 days, how many bars will he eat a day?

10 points

⟨Ans.⟩ _____

10 We are making flash cards to study for our test. We can make 8 flash cards out of 1 piece of paper. If we divide 24 sheets of paper evenly among our group of 3, how many cards will each of us make?

10 points

⟨Ans.⟩ _____

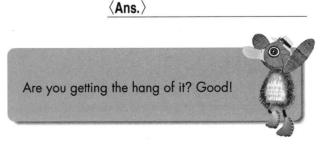

Are you getting the hang of it? Good!

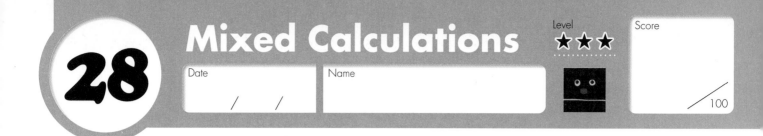
1 It is back-to-school time, and the art teacher bought 2 sets of glass beads and 3 sets of ceramic beads. If each set of glass beads has 40 beads in each, and each set of ceramic beads has 80 beads in each, how many beads did she buy in total?

10 points

Total glass beads

Total ceramic beads

Total beads

⟨Ans.⟩

2 The farmer was selling fruit at the market. He was selling 4 crates of 90 peaches each and 8 crates of 30 mangos each. How many pieces of fruit was he selling?

10 points

⟨Ans.⟩

3 Mr. Elmore bought 2 ties for $5 each, and 5 handkerchiefs for $3 each. How much did he spend in all?

10 points

⟨Ans.⟩

4 For gym class today, the boys formed 4 lines with 15 boys each, and the girls formed 4 lines with 12 girls each. How many people were in the class in all?

10 points

⟨Ans.⟩

5 The principal bought breakfast for her school today. She bought 4 boxes of 80 bagels and 6 boxes of 40 doughnuts. How many more bagels did she buy?

15 points

⟨Ans.⟩ _____

6 You arranged some stones as shown below. How many more black stones than white stones are there?

15 points

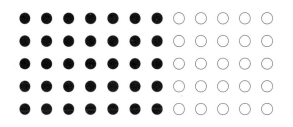

⟨Ans.⟩ _____

7 Veronica and her sister went shopping for hair clips. One hair clip is 8¢, and Veronica's sister bought 32¢ worth. If Veronica spend 48¢, how many more clips did she buy than her sister?

15 points

⟨Ans.⟩ _____

8 Bobby is looking at a 72¢ sticker set that contains 9 stickers. He also found a set that costs 36¢ and has 6 stickers. What is the difference in price per sticker?

15 points

⟨Ans.⟩ _____

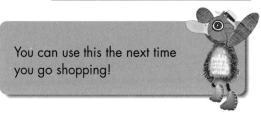

You can use this the next time you go shopping!

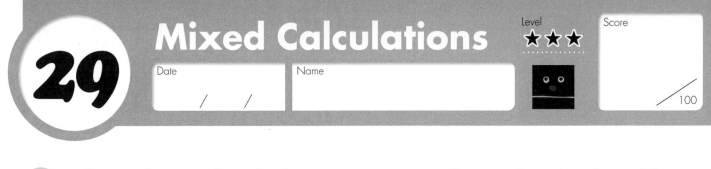

Mixed Calculations

Level ★★★

Date   /   /

Name

Score
/ 100

1 The gardener at the school is experimenting with trees. First, he planted 4 trees at 10-foot intervals in a line. How long was it from the first tree to the last tree?

10 points

$$10 \times 3 =$$

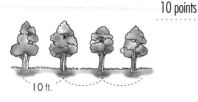

10 ft.

〈Ans.〉 _____

2 Then, he decided to plant 5 trees at 10-foot intervals in a line. How far was it from the first tree to the last tree then?

10 points

〈Ans.〉 _____

3 He still didn't like it, so he planted 6 trees at 10-foot intervals. How long was the line of trees then?

10 points

〈Ans.〉 _____

4 He finally settled on 7 trees at 10-foot intervals. How far was it from the first tree to the last tree when he was done?

10 points

〈Ans.〉 _____

5 For the 4th of July, Mr. Joseph always puts out 9 flags in a straight line. If he puts them at 8-foot intervals, how far is from the first flag to the last flag?

10 points

〈Ans.〉 _____

6 Around the pond at Mr. Pope's workplace, there are 5 trees in intervals of 10 meters. How far is it around the pond?

10 points

$10 \times 5 =$

⟨Ans.⟩ _____

7 Mother planted 8 bushes at 10-meter intervals around our back yard. How far is it around our back yard?

10 points

⟨Ans.⟩ _____

8 There are 10 trees in 15-meter intervals around the pond in our park. How far is it around the pond?

10 points

⟨Ans.⟩ _____

9 Mrs. Kwan put 20 pegs in 3-meter intervals around her flowerbed. How far is it around her flowerbed?

10 points

⟨Ans.⟩ _____

10 In the small park in our town, there are 15 trees in 12-meter intervals around the park. How far is it around the park?

10 points

⟨Ans.⟩ _____

Keep up the good work!

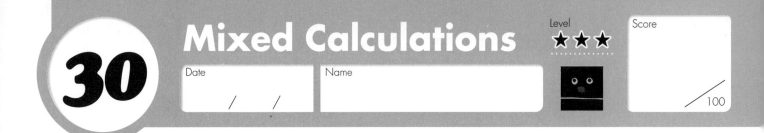

Mixed Calculations

Level ★★★

Date / /

Name

Score /100

1 You connected 2 pieces of tape as shown below. If each piece was 100 inches long, how long is the new tape from end to end?

15 points

☐ in.
10 in.

100 in.   100 in.

$100 \times 2 = 200$

$200 - 10 =$

⟨Ans.⟩ _____

2 You connected 2 pieces of tape that were 100 inches long as shown below. How long would the new piece be from end to end?

15 points

☐ in.
20 in.

100 in.   100 in.

⟨Ans.⟩ _____

3 You connected 2 pieces of tape that were 150 inches long as shown below. How long would the new piece be from end to end?

15 points

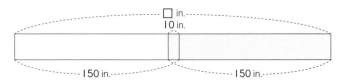

☐ in.
10 in.

150 in.   150 in.

⟨Ans.⟩ _____

4 Jessica was making a craft and had only 2 pieces of tape that were 150 centimeters long. She combined them as shown below. How long was her new piece of tape?

15 points

☐ cm
20 cm

----150 cm----   ----150 cm----

⟨**Ans.**⟩ _____

5 Mr. Ahn doesn't like to keep tape around if it is almost done. He connects them together instead so that he can keep one longer tape around. If he connected 2 pieces of tape that were 200 centimeters long as shown below, how long would his new piece of tape be?

20 points

☐ cm
10 cm

----200 cm----   ----200 cm----

⟨**Ans.**⟩ _____

6 Mr. Ahn's tape broke apart, so he re-attached the 2 pieces of tape as shown below. If each piece was 200 centimeters long, how long is his new piece of tape?

20 points

☐ cm
25 cm

----200 cm----   ----200 cm----

⟨**Ans.**⟩ _____

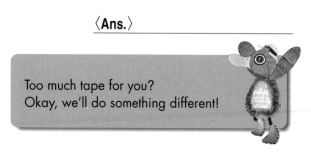

Too much tape for you?
Okay, we'll do something different!

31 Mixed Calculations

Level ★★★

Date / /

Name

Score

/100

1 Mrs. Henry has a small garden with some red flowers and some white flowers. There are 11 total flowers right now. If there is 1 more red flower than white flowers, how many white flowers does she have in the garden? 10 points

$$11 - 1 = 10$$

$$10 \div 2 =$$

Red ⎫
White ⎬ Total is 11
1

⟨Ans.⟩ _____

2 Now, Mrs. Henry has 12 flowers. If there are 2 more red flowers than white flowers, how many white flowers does she have now? 10 points

$$12 - 2 = 10$$

$$10 \div 2 =$$

Red ⎫
White ⎬ Total is 12
2

⟨Ans.⟩ _____

3 Some more flowers grew in Mrs. Henry's garden, and now she has 15 flowers. If there are 3 more red flowers than white flowers, how many white flowers does she have now? 10 points

Red ⎫
White ⎬ Total is 15
3

⟨Ans.⟩ _____

4 There are 24 red and white gumballs in Jessie's gumball machine. If there are 6 more red gumballs than white gumballs, how many white gumballs does Jessie have? 10 points

Red ⎫
White ⎬ Total is 24
6

⟨Ans.⟩ _____

5 Rudy got a bag of mixed nuts for the holidays. There are 30 total nuts in the bag. He noticed that he had 14 more chestnuts than walnuts. How many walnuts are there? 10 points

⟨Ans.⟩ _____

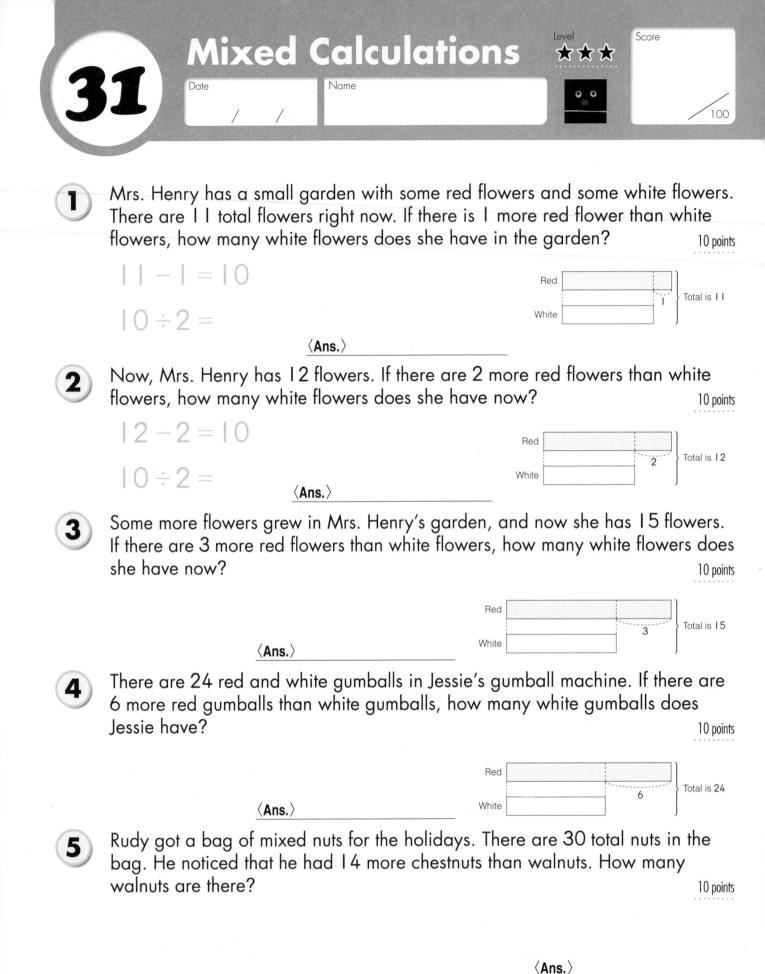

6 We are playing tag today with 11 people. If there is 1 more boy than there are girls, how many boys are playing tag?

10 points

$11 + 1 = 12$

$12 \div 2 =$

Boys		1
Girls		

Total is 11

⟨Ans.⟩ _____

7 In your class today, you have 12 people, and there are 2 more boys than girls. How many boys are there in class today?

10 points

$12 + 2 = 14$

$14 \div 2 =$

	2
Boys	
Girls	

Total is 12

⟨Ans.⟩ _____

8 At the doctor's office today, there are 15 girls and boys waiting their turn. If there are 3 more boys than girls, how many boys are at the doctor's office?

10 points

	3
Boys	
Girls	

Total is 15

⟨Ans.⟩ _____

9 John really likes blue and white shirts, and has 14 of them in all. If he has 2 more white shirts than blue shirts, how many white shirts does he have?

10 points

⟨Ans.⟩ _____

10 We got a mixed box of chocolates over the holidays. It had 3 more dark chocolate pieces than milk chocolate. If there were 15 pieces total, how many dark chocolates were in the box?

10 points

⟨Ans.⟩ _____

That was tough. Well done!

Graphs & Tables

1 When you asked everyone in the class what their favorite fruit was, you got the result shown below.

ⓐ melon	ⓕ pear	ⓚ peach	ⓟ melon	ⓤ orange
ⓑ apple	ⓖ orange	ⓛ grapes	ⓠ orange	ⓥ melon
ⓒ grapes	ⓗ melon	ⓜ apple	ⓡ melon	ⓦ grapes
ⓓ orange	ⓘ melon	ⓝ melon	ⓢ apple	ⓧ orange
ⓔ melon	ⓙ apple	ⓞ orange	ⓣ orange	ⓨ melon

(1) Use the table below to tally everyone's favorite fruit. Use hash marks as shown to the right. As you can see, ⓐ through ⓔ have already been counted in the table, so start with ⓕ.

20 points for completion

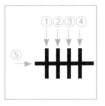

Favorite fruit

grapes	apple	melon	orange	pear	peach
I	I	II	I		

(2) Rewrite the results of (1) in the table below. Use numbers instead of hash marks.

10 points for completion

Favorite fruit

Fruit	grapes	apple	melon	orange	the others	Total
Number of people						

2 The graph on the right is called a *bar graph*. Answer the questions below using this bar graph.

10 points per question

(1) How many students does each vertical box represent?

(　　　)

(2) How many students like each type of book?

novel (　　　) comic book (　　　)

photo book (　　　) history book (　　　)

(3) How many more students like novels than comic books?

(　　　)

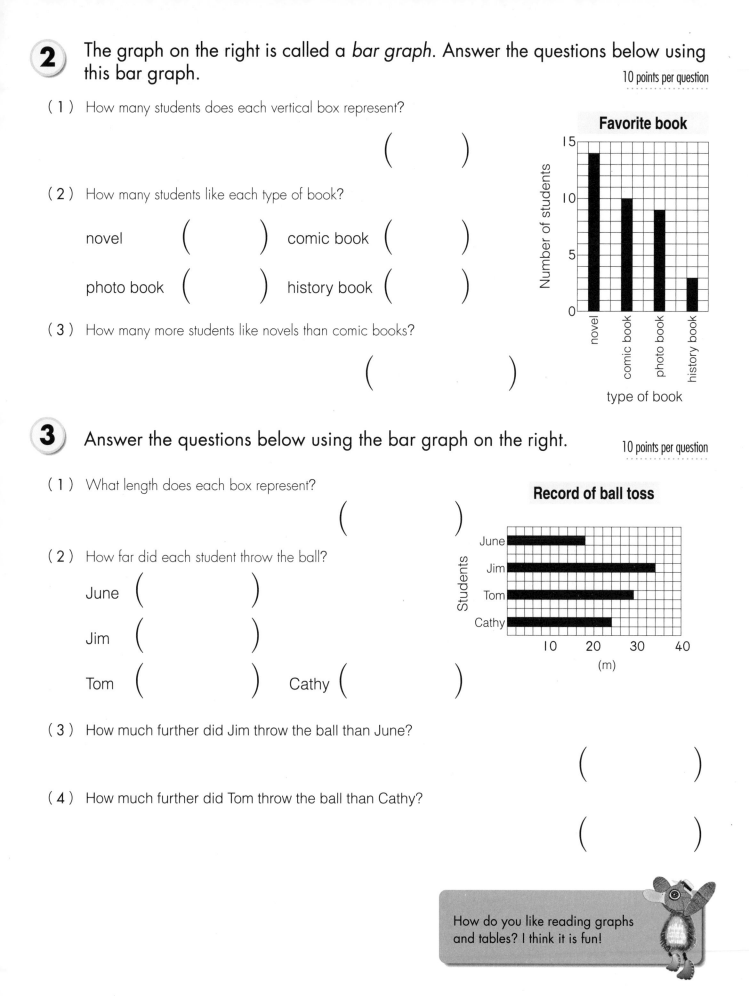

Favorite book

3 Answer the questions below using the bar graph on the right.

10 points per question

(1) What length does each box represent?

(　　　)

(2) How far did each student throw the ball?

June (　　　)

Jim (　　　)

Tom (　　　) Cathy (　　　)

(3) How much further did Jim throw the ball than June?

(　　　)

(4) How much further did Tom throw the ball than Cathy?

(　　　)

Record of ball toss

How do you like reading graphs and tables? I think it is fun!

Graphs & Tables

Level ★★

Date   /   /

Name

Score
/100

1 You asked your classmates what their favorite color was, and made the table on the left. In order to draw the bar graph below, follow questions (1) through (4).

Favorite color

Color	Number of people
Red	10
Blue	8
Yellow	11
Green	12
Brown	7

(1) Write the remaining color names in **A**-**C**.

15 points for completion

(2) Write the appropriate numbers in **D** and **E**. Write the units in the ( ) at the top.

10 points for completion

(3) Using the head-count, draw the appropriate bars for each color.

25 points for completion

(4) Write the title of the graph in **F**.

5 points

(5) What color has the longest bar?

5 points

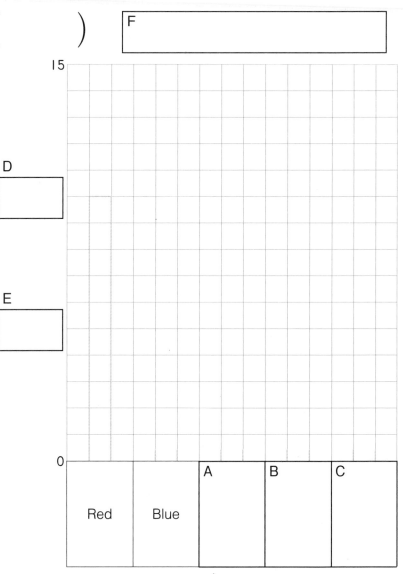

2 You asked all your classmates what their favorite sport was this time. Using the table you made on the right, draw the bar graph below.

10 points per question

(1) Write the names of the sports in the boxes at the bottom.

(2) Write the unit numbers along the left side of the graph, and the units in the ( ) at the top.

(3) Using your head-count table, draw the appropriate bars for each sport.

(4) Write the title of the graph.

Favorite sport

Sports	Number of people
Baseball	19
Basketball	15
Swimming	16
Football	13
Running	7

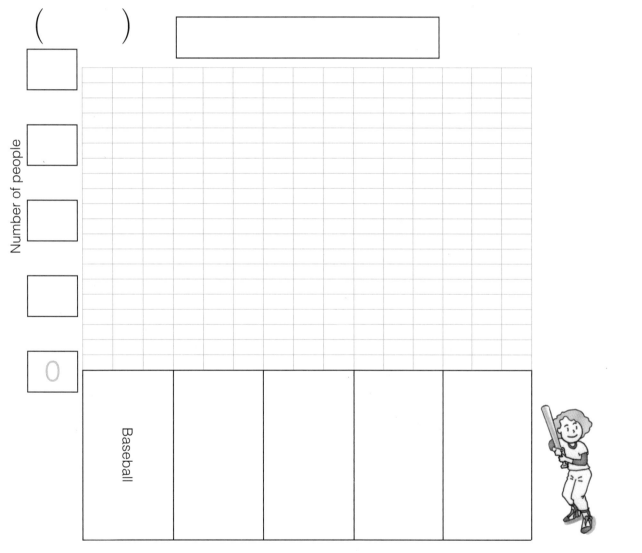

( )

Number of people

0

Baseball

sport

That was fun, right?

1 The school nurse wrote down the number of people missing from class in classes A, B and C. She put the numbers of absentees in the tables below.

7 points per question

Absent from class A

Month	Number of people
April	25
May	13
June	18
July	8
Total	ⓐ

Absent from class B

Month	Number of people
April	21
May	10
June	19
July	11
Total	ⓑ

Absent from class C

Month	Number of people
April	19
May	17
June	9
July	10
Total	ⓒ

(1) How many people missed class in all three classes from April to July? Write the sum of ⓐ, ⓑ and ⓒ on each table.

(2) Gather all three tables into one single graph below.

(3) Total all the absentees for each month and write the sum in ⓟ to ⓢ on the table.

(4) In ⓣ on the table, write the total number of absentees from April to July in Grade 3.

Absent from Grade 3

Class / Month	A	B	C	Total
April	25	ⓕ	ⓚ	ⓟ 65
May	13	ⓖ	ⓛ	ⓠ
June	18	ⓗ	ⓜ	ⓡ
July	ⓓ 8	ⓘ	ⓝ	ⓢ
Total	ⓔ 64	ⓙ	ⓞ	ⓣ

© Kumon Publishing Co., Ltd.

2 Students in classes A, B and C selected their favorite types of books, and you tallied it in the table below.

9 points per question

Favorite type of book

Class / Book	A	B	C	Total
Novel	14	12	10	ⓓ
Comic book	10	11	13	ⓔ
Photo book	9	7	10	ⓕ
Biography	3	5	4	ⓖ
Total	ⓐ	ⓑ	ⓒ	ⓗ

(1) How many students in class B said a comic book was their favorite type of book?

(   )

(2) How many students in class C said a biography was their favorite type of book?

(   )

(3) How many students were in each class? Write the sums in ⓐ-ⓒ.

(4) Which class has the fewest students?

(   )

(5) Write the sum of each type of book from ⓓ to ⓖ.

(6) Which type of book is favored by the most students?

(   )

(7) Write the appropriate number in ⓗ.

(8) What does the number in ⓗ represent?

(   )

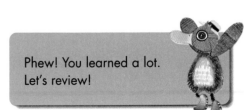

Phew! You learned a lot. Let's review!

35

Review

Date
/ /

Name

Level
★ ★ ★

Score
/100

1 There are 236 children and 187 adults at the amusement park today.

5 points per question

(1) How many people are there in all?

⟨Ans.⟩ _____

(2) How many more children than adults are there?

⟨Ans.⟩ _____

2 The distance from Tim's house to the train station is 450 meters. The distance from the train station to city hall is 380 meters. If Tim goes past the train station, how far is it from his house to city hall?

10 points

⟨Ans.⟩ _____

3 Kim went to the library and came back. She took 8 minutes to get to the library and 13 minutes to come back. How long did it take her in all?

10 points

⟨Ans.⟩ _____

4 Farmer Brown put 650 grams of wheat into a box that weighs 260 grams. How much does the box weigh now?

10 points

⟨Ans.⟩ _____

5 Peter had to carry some concrete blocks to the town dump in his car. If he could fit 12 blocks in his car each time, and had to go 15 times, how many blocks did he carry in all?

10 points

⟨Ans.⟩ _____

6 Irene is making a craft and cut 5 pieces of tape. Each length was 24 centimeters, and there were 60 centimeters remaining. How long was the tape at first?

10 points

〈Ans.〉 _____

7 The hens laid 63 eggs today. If you put 9 eggs in each box, how many boxes will you need?

10 points

〈Ans.〉 _____

8 The Tang family got 54 candies in the mail over the holidays. If all 8 children get the same number of candies, how many does each child get? And how many candies remain?

10 points

〈Ans.〉 _____

9 We picked 63 oranges off the trees today. If we put 8 oranges into each bag, how many bags will we need? How many oranges will be left over?

10 points

〈Ans.〉 _____

10 There are 22 third-graders in Sally's town. If there are 4 more boys than girls, how many girls are there?

10 points

〈Ans.〉 _____

Almost there!

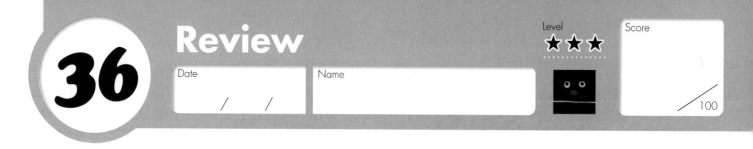

1 You bought 25 bouquets, and each bouquet had 30 flowers in it. How many flowers did you buy?

10 points

⟨Ans.⟩ _____

2 The distance from Lily's house to the bus station is 950 meters, and from her house to the store it is 670 meters. How much further is it to the bus station than to the store?

10 points

950 m

670 m

⟨Ans.⟩ _____

3 Bob has a small bucket, a big bucket, and a tank. He can fit 2 liters of water in the small bucket. The big bucket can carry 2 times as much water as the small bucket, and the tank can carry 3 times as much water as the big bucket. How much water can the tank contain?

10 points

⟨Ans.⟩ _____

4 You weighed some water with a 350-gram bucket, and the total weight was 920 grams. How much did the water weigh?

10 points

⟨Ans.⟩ _____

5 Mother bought 24 oranges, and each weighed 75 grams. How much did the oranges weigh in all?

10 points

⟨Ans.⟩ _____

6 A big can of olives has 85 olives in it. If Tammy buys 36 cans, how many olives will she have?

10 points

〈Ans.〉 _____

7 Jane is looking at a set of 8 hair clips that cost 56¢. How much is each hair clip?

10 points

〈Ans.〉 _____

8 Grandfather is making bunches out of his bamboo sticks. If he has 40 bamboo sticks, and 6 sticks go in a bunch, how many bunches can he make?

10 points

〈Ans.〉 _____

9 It takes 45 minutes for Ted to get to the train station from his house every morning. If he wants to arrive at the station at 9 in the morning, what time does he have to leave his house?

10 points

〈Ans.〉 _____

10 There are 15 trees at intervals of 6 meters along my street. How far is it from the first tree to the last tree?

10 points

〈Ans.〉 _____

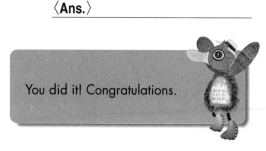

You did it! Congratulations.

1 Review
pp 2,3

1 46 − 7 = 39 **Ans.** 39 times.
2 65 − 35 = 30 **Ans.** 30 jugs
3 2 m 35 cm − 1 m 70 cm = 65 cm
  Ans. 65 cm
4 20 − 12 = 8 **Ans.** 8 candies
5 7 + 9 = 16, 40 − 16 = 24
  [Also, 40 − 7 = 33, 33 − 9 = 24]
  Ans. 24 stickers
6 93 − 45 = 48 **Ans.** 48 pages
7 90 − 40 = 50 **Ans.** 50¢
8 35 − 6 = 29, 29 + 4 = 33 **Ans.** 33 pigeons
9 26 − 25 = 1 **Ans.** 1 candy
10 10 ft. 10 in. + 15 ft. 1 in. = 25 ft. 11 in.
  Ans. 25 ft. 11 in.

2 Review
pp 4,5

1 90 − 65 = 25 **Ans.** 25 notebooks
2 32 − 6 = 26 **Ans.** 26 flowers
3 4 ft. 5 in. + 6 in. = 4 ft. 11 in.
  Ans. 4 ft. 11 in.
4 28 + 22 = 50 **Ans.** 50 problems
5 57 + 13 = 70 **Ans.** 70¢
6 38 − 18 = 20 **Ans.** 20 people
7 95 − 46 = 49 **Ans.** 49 adults
8 32 + 14 = 46 **Ans.** 46 bags
9 5 + 10 = 15 **Ans.** 15 m
10 18 + 10 − 25 = 3 **Ans.** 3 sheets

3 Addition & Subtraction
pp 6,7

1 240 + 250 = 490 **Ans.** 490 mL
2 450 + 85 = 535 **Ans.** 535 g
3 135 + 122 = 257 **Ans.** 257 flowers
4 450 + 680 = 1,130 **Ans.** 1,130 cm
5 178 + 169 = 347 **Ans.** 347 students
6 245 + 178 = 423 **Ans.** 423 children
7 283 + 175 = 458 **Ans.** 458 books
8 359 + 486 = 845 **Ans.** 845 cranes
9 128 + 147 = 275 **Ans.** 275 strawberries
10 165 + 187 = 352 **Ans.** 352 pieces

4 Addition & Subtraction
pp 8,9

1 770 − 250 = 520 **Ans.** 520 mL
2 185 − 133 = 52 **Ans.** 52 red flowers
3 128 − 75 = 53 **Ans.** 53 stickers
4 178 − 169 = 9 **Ans.** There are 9 more boys.
5 800 − 245 = 555 **Ans.** 555 yd.
6 240 − 83 = 157 **Ans.** 157 pages
7 564 − 296 = 268 **Ans.** 268 girls
8 500 − 265 = 235 **Ans.** 235 sheets
9 562 − 175 = 387 **Ans.** 387 students
10 324 − 143 = 181 **Ans.** 181 adults

5 Addition & Subtraction
pp 10,11

1 264 − 86 = 178 **Ans.** 178 people
2 175 + 179 = 354 **Ans.** 354 people
3 548 + 497 = 1,045 **Ans.** 1,045 people
4 500 − 348 = 152 **Ans.** 152 pennies
5 125 + 175 = 300 **Ans.** 300 stickers
6 240 − 176 = 64 **Ans.** 64 pages
7 365 − 196 = 169 **Ans.** 169 girls
8 500 + 265 = 765 **Ans.** 765 in.
9 360 + 250 = 610 **Ans.** 610 sticks
10 263 − 185 = 78 **Ans.** 78 oranges

6 Length
pp 12,13

1 300 mi. + 500 mi. = 800 mi. **Ans.** 800 mi.
2 400 ft. + 200 ft. = 600 ft. **Ans.** 600 ft.
3 500 m − 300 m = 200 m **Ans.** 200 m
4 400 mi. − 200 mi. = 200 mi. **Ans.** 200 mi.

5 (1) 700 m + 300 m = 1,000 m   **Ans.**   1,000 m

   (2) 800 m − 500 m = 300 m   **Ans.**   300 m

6 4 km 500 m + 1 km 200 m = 5 km 700 m

                                Ans.   5 km 700 m

7 800 m + 800 m = 1 km 600 m

                                Ans.   1 km 600 m

8 2 km − 1 km 200 m = 800 m   **Ans.**   800 m

7 Time   pp 14, 15

1   **Ans.**   16 minutes

2   **Ans.**   28 minutes

3   **Ans.**   30 minutes

4   **Ans.**   25 minutes

5   **Ans.**   40 minutes

6   **Ans.**   1 hour

7   **Ans.**   2 hours

8   **Ans.**   3 hours

9   **Ans.**   4 hours

10   **Ans.**   1 hour

8 Time   pp 16, 17

1   **Ans.**   8 : 15

2   **Ans.**   4 : 45

3   **Ans.**   3 : 50

4   **Ans.**   10 : 55

5   **Ans.**   10 : 00

6   **Ans.**   8 : 50

7   **Ans.**   9 : 45

8   **Ans.**   4 : 20

9   **Ans.**   4 : 00

10   **Ans.**   3 : 10

9 Multiplication   pp 18, 19

1 2 × 4 = 8   **Ans.**   8 oranges

2 2 × 8 = 16   **Ans.**   16 children

3 3 × 4 = 12   **Ans.**   12 flowers

4 3 × 5 = 15   **Ans.**   15 people

5 2 × 9 = 18   **Ans.**   18 pencils

6 3 × 7 = 21   **Ans.**   21 flowers

7 2 × 3 = 6   **Ans.**   6 people

8 3 × 2 = 6   **Ans.**   6 cm

9 2 × 7 = 14   **Ans.**   14 pills

10 3 × 6 = 18   **Ans.**   18 ft.

10 Multiplication   pp 20, 21

1 4 × 6 = 24   **Ans.**   24 people

2 5 × 7 = 35   **Ans.**   35 oranges

3 4 × 5 = 20   **Ans.**   20 candies

4 4 × 9 = 36   **Ans.**   36 sheets

5 5 × 6 = 30   **Ans.**   30 in.

6 5 × 3 = 15   **Ans.**   15 pencils

7 5 × 3 = 15   **Ans.**   15 apples

8 4 × 5 = 20   **Ans.**   20 bottles

9 5 × 8 = 40   **Ans.**   40 pencils

10 4 × 3 = 12   **Ans.**   12 m

11 Multiplication   pp 22, 23

1 6 × 4 = 24   **Ans.**   24 candies

2 6 × 3 = 18   **Ans.**   18 cm

3 7 × 5 = 35   **Ans.**   35 coins

4 7 × 4 = 28   **Ans.**   28 pens

5 7 × 3 = 21   **Ans.**   21 days

6 6 × 5 = 30   **Ans.**   30 people

7 6 × 5 = 30   **Ans.**   30 people

8 7 × 6 = 42   **Ans.**   42 pieces

9 6 × 9 = 54   **Ans.**   54 bottles

10 8 × 7 = 56   **Ans.**   56 cm

12 Multiplication   pp 24, 25

1 8 × 5 = 40   **Ans.**   40 chocolates

2 9 × 7 = 63   **Ans.**   63 players

3 8 × 3 = 24   **Ans.**   24 flowers

4 8 × 6 = 48   **Ans.**   48 m

5 9 × 4 = 36   **Ans.**   36 cm

6 8 × 7 = 56   **Ans.**   56 cm

7 9 × 5 = 45   **Ans.**   45¢

8 8 × 4 = 32   **Ans.**   32 students

9 8 × 9 = 72   **Ans.**   72 people

(10) $9 \times 3 = 27$ **Ans.**   27 pieces

(13) Multiplication
pp 26, 27

(1) $6 \times 6 = 36$ **Ans.**   36 people
(2) $8 \times 4 = 32$ **Ans.**   32 chestnuts
(3) $6 \times 7 = 42$ **Ans.**   42 pencils
(4) $5 \times 8 = 40$ **Ans.**   40¢
(5) $6 \times 3 = 18$ **Ans.**   18 pieces
(6) $4 \times 5 = 20$ **Ans.**   20 triangles
(7) $7 \times 8 = 56$ **Ans.**   56 cm
(8) $5 \times 9 = 45$ **Ans.**   45 people
(9) $3 \times 7 = 21$ **Ans.**   21 m
(10) $7 \times 9 = 63$ **Ans.**   63 flowers

(14) Multiplication
pp 28, 29

(1) $30 \times 4 = 120$ **Ans.**   120 sheets
(2) $60 \times 3 = 180$ **Ans.**   180 pencils
(3) $50 \times 4 = 200$ **Ans.**   200 stamps
(4) $4 \times 10 = 40$ **Ans.**   40 oranges
(5) $5 \times 10 = 50$ **Ans.**   50 people
(6) $8 \times 30 = 240$ **Ans.**   240 candies
(7) $6 \times 40 = 240$ **Ans.**   240 flowers
(8) $5 \times 32 = 160$ **Ans.**   160 seeds
(9) $20 \times 18 = 360$ **Ans.**   360 crayons
(10) $3 \times 36 = 108$ **Ans.**   108 sheets

(15) Multiplication
pp 30, 31

(1) $4 \times 2 \times 5 = 40$ **Ans.**   40 candies
(2) $5 \times 3 \times 3 = 45$ **Ans.**   45 oranges
(3) $70 \times 5 \times 2 = 700$ **Ans.**   700 pretzels
(4) $45 \times 2 \times 3 = 270$ **Ans.**   270 stickers
(5) $3 \times 2 \times 4 = 24$ **Ans.**   24 cakes
(6) $8 \times 2 \times 3 = 48$ **Ans.**   48 cm
(7) $2 \times 3 \times 5 = 30$ **Ans.**   30 pills
(8) $3 \times 3 \times 4 = 36$ **Ans.**   36 pills

(16) Division
pp 32, 33

(1) $6 \div 2 = 3$ **Ans.**   3 candies
(2) $8 \div 2 = 4$ **Ans.**   4 sheets

(3) $6 \div 3 = 2$ **Ans.**   2 pencils
(4) $12 \div 3 = 4$ **Ans.**   4 candies
(5) $15 \div 3 = 5$ **Ans.**   5 pencils
(6) $15 \div 5 = 3$ **Ans.**   3 pencils
(7) $18 \div 3 = 6$ **Ans.**   6 chocolates
(8) $24 \div 4 = 6$ **Ans.**   6 sheets
(9) $25 \div 5 = 5$ **Ans.**   5 cm
(10) $28 \div 7 = 4$ **Ans.**   4 oranges

(17) Division
pp 34, 35

(1) $6 \div 2 = 3$ **Ans.**   3 people
(2) $8 \div 2 = 4$ **Ans.**   4 people
(3) $6 \div 3 = 2$ **Ans.**   2 people
(4) $12 \div 3 = 4$ **Ans.**   4 plates
(5) $12 \div 4 = 3$ **Ans.**   3 friends
(6) $15 \div 3 = 5$ **Ans.**   5 friends
(7) $15 \div 5 = 3$ **Ans.**   3 friends
(8) $18 \div 3 = 6$ **Ans.**   6 people
(9) $24 \div 6 = 4$ **Ans.**   4 vases
(10) $32 \div 8 = 4$ **Ans.**   4 pieces

(18) Division
pp 36, 37

(1) $21 \div 3 = 7$ **Ans.**   7 pencils
(2) $21 \div 3 = 7$ **Ans.**   7 people
(3) $30 \div 5 = 6$ **Ans.**   6 L
(4) $40 \div 8 = 5$ **Ans.**   5¢
(5) $72 \div 8 = 9$ **Ans.**   9 packs
(6) $40 \div 8 = 5$ **Ans.**   5 soldiers
(7) $30 \div 5 = 6$ **Ans.**   6 neighbors
(8) $20 \div 5 = 4$ **Ans.**   4 days
(9) $18 \div 2 = 9$ **Ans.**   9 bottles
(10) 5 cm 4 mm = 54 mm,
$54 \div 6 = 9$ **Ans.**   9 mm

(19) Division
pp 38, 39

(1) $20 \div 3 = 6$ R 2 **Ans.**   2 coins remain.
(2) $30 \div 4 = 7$ R 2 **Ans.**   2 sheets remain.

(3) $27 \div 6 = 4 \text{ R } 3$

Ans. 4 oranges in each bag, 3 oranges remain.

(4) $40 \div 7 = 5 \text{ R } 5$

Ans. 5 pieces per child, 5 pieces remain.

(5) $27 \div 6 = 4 \text{ R } 3$

Ans. 4 fish per bucket, 3 fish remain.

(6) $20 \div 3 = 6 \text{ R } 2$ **Ans.** 2 slices remain

(7) $30 \div 4 = 7 \text{ R } 2$ **Ans.** 2 treats remain

(8) $36 \div 8 = 4 \text{ R } 4$

Ans. 4 bags, 4 cakes will remain.

(9) $75 \div 8 = 9 \text{ R } 3$

Ans. 9 presents, 3 inches remain.

(10) $65 \div 7 = 9 \text{ R } 2$

Ans. 9 baskets, 2 apples remain.

20 Division
pp 40,41

(1) $45 \div 6 = 7 \text{ R } 3$

Ans. 7 people, 3 marbles remain.

(2) $60 \div 9 = 6 \text{ R } 6$

Ans. 6 vases, 6 carnations remain.

(3) $38 \div 8 = 4 \text{ R } 6$

Ans. 4 sheets each, 6 sheets remain.

(4) $60 \div 7 = 8 \text{ R } 4$

Ans. 8 cranes each, 4 sheets remain.

(5) $38 \div 6 = 6 \text{ R } 2$

Ans. 6 tanks needed, 2 turtles remain.

(6) $25 \div 4 = 6 \text{ R } 1$ **Ans.** 6 pictures

(7) $50 \div 6 = 8 \text{ R } 2$ **Ans.** 8 books

(8) $26 \div 4 = 6 \text{ R } 2$ **Ans.** 7 boats

(9) $30 \div 8 = 3 \text{ R } 6$ **Ans.** 4 sheets

(10) $50 \div 6 = 8 \text{ R } 2$ **Ans.** 9 containers

21 Mixed Calculations
pp 42,43

(1) (1) $6 + 7 = 13$ **Ans.** 13 cars

(2) $15 + 13 = 28$ **Ans.** 28 cars

(2) $4 + 8 = 12$, $15 + 12 = 27$ **Ans.** 27 cookies

(3) $5 + 7 = 12$, $12 + 12 = 24$ **Ans.** 24 sparrows

(4) $14 + 3 = 17$, $9 + 17 = 26$ **Ans.** 26 geese

(5) $8 + 11 = 19$, $18 + 19 = 37$ **Ans.** 37 slices

(6) $15 + 10 = 25$, $35 + 25 = 60$

Ans. 60 sheets

(7) $11 + 13 = 24$, $12 + 24 = 36$

Ans. 36 people

(8) $50 + 40 = 90$, $130 + 90 = 220$

Ans. 220 inches

22 Mixed Calculations
pp 44,45

(1) (1) $60 + 80 = 140$

Ans. 140 blueberries and raspberries

(2) $210 - 140 = 70$ **Ans.** 70 blackberries

(2) (1) $70 + 40 = 110$

Ans. 110 crayons and markers

(2) $230 - 110 = 120$ **Ans.** 120 paintbrushes

(3) $36 + 25 = 61$, $82 - 61 = 21$

Ans. 21 blue sheets

(4) $80 + 75 = 155$, $270 - 155 = 115$

Ans. 115 candies

(5) $14 + 13 = 27$, $33 - 27 = 6$ **Ans.** 6 girls

(6) $18 + 23 = 41$, $60 - 41 = 19$ **Ans.** 19 stickers

(7) $16 + 15 = 31$, $48 - 31 = 17$ **Ans.** 17 sheets

23 Mixed Calculations
pp 46,47

(1) (1) $5 \times 4 = 20$ **Ans.** 20 stickers

(2) $30 + 20 = 50$ **Ans.** 50 stickers

(2) (1) $6 \times 7 = 42$ **Ans.** 42 balloons

(2) $18 + 42 = 60$ **Ans.** 60 balloons

(3) $2 \times 10 = 20$, $3 + 20 = 23$ **Ans.** 23 bulbs

(4) $4 \times 8 = 32$, $21 + 32 = 53$ **Ans.** 53 coupons

(5) $20 \times 7 = 140$, $60 + 140 = 200$

Ans. 200 carrots

(6) $25 \times 6 = 150$, $150 - 8 = 142$

Ans. 142 logs

(7) $\$5 = 500$¢, $76 \times 6 = 456$,
$500 - 456 = 44$ **Ans.** 44 ¢

(8) $\$2 = 200$¢, $65 \times 3 = 195$,
$200 - 195 = 5$ **Ans.** 5¢

24 Mixed Calculations pp 48,49

1 (1) $27 \div 3 = 9$ **Ans.** 9 dishes
   (2) $5 + 9 = 14$ **Ans.** 14 dishes

2 (1) $16 \div 2 = 8$ **Ans.** 8 pots
   (2) $4 + 8 = 12$ **Ans.** 12 pots

3 $32 \div 4 = 8,\ 5 + 8 = 13$ **Ans.** 13 children

4 $72 \div 8 = 9,\ 9 - 2 = 7$ **Ans.** 7 boxes

5 $54 \div 6 = 9,\ 12 - 9 = 3$ **Ans.** 3 baskets

6 $18 \div 2 = 9,\ 25 + 9 = 34$ **Ans.** 34 stickers

7 $21 \div 3 = 7,\ 36 + 7 = 43$ **Ans.** 43 pennies

25 Mixed Calculations pp 50,51

1 $15 - 6 = 9$ **Ans.** 9 children

2 (1) $8 + 7 = 15$ **Ans.** 15 children
   (2) $15 - 6 = 9$ **Ans.** 9 children

3 $15 + 7 = 22,\ 22 - 12 = 10$ **Ans.** 10 children

4 $8 + 12 = 20,\ 20 - 7 = 13$ **Ans.** 13 children

5 $13 + 9 = 22,\ 22 - 8 = 14$ **Ans.** 14 flowers

6 $22 + 15 = 37,\ 37 - 19 = 18$ **Ans.** 18 candies

7 $17 + 7 = 24,\ 24 - 11 = 13$
          Ans. 13 ducks and swans

26 Mixed Calculations pp 52,53

1 $24 \div 6 = 4$ **Ans.** 4 strawberries

2 (1) $3 + 2 = 5$ **Ans.** 5 children
   (2) $25 \div 5 = 5$ **Ans.** 5 oranges

3 $3 + 4 = 7,\ 21 \div 7 = 3$ **Ans.** 3 sheets

4 $4 + 2 = 6,\ 30 \div 6 = 5$ **Ans.** 5 oranges

5 (1) $2 \times 4 = 8$ **Ans.** 8 candies
   (2) $40 \div 8 = 5$ **Ans.** 5¢

6 $3 \times 2 = 6,\ 42 \div 6 = 7$ **Ans.** 7¢

7 $2 \times 3 = 6,\ 48 \div 6 = 8$ **Ans.** 8 peanuts

8 $3 \times 3 = 9,\ 45 \div 9 = 5$ **Ans.** 5¢

27 Mixed Calculations pp 54,55

1 $6 \times 4 = 24,\ 24 \div 8 = 3$ **Ans.** 3 oranges

2 $12 \times 3 = 36,\ 36 \div 6 = 6$ **Ans.** 6 pencils

3 $8 \times 3 = 24,\ 24 \div 4 = 6$ **Ans.** 6 girls

---

4 $36 \times 2 = 72,\ 72 \div 8 = 9$ **Ans.** 9 cranes

5 $6 \times 4 = 24,\ 24 \div 8 = 3$ **Ans.** 3 bunches

6 $27 \div 3 = 9,\ 9 \times 5 = 45$ **Ans.** 45¢

7 $24 \div 3 = 8,\ 8 \times 7 = 56$ **Ans.** 56 g

8 $45 \div 5 = 9,\ 3 \times 9 = 27$ **Ans.** $27

9 $18 \div 2 = 9,\ 9 \div 3 = 3$ **Ans.** 3 bars

10 $24 \div 3 = 8,\ 8 \times 8 = 64$ **Ans.** 64 cards
[Also, $8 \times 24 = 192,\ 192 \div 3 = 64$]

28 Mixed Calculations pp 56,57

1 $40 \times 2 = 80,\ 80 \times 3 = 240,$
  $80 + 240 = 320$ **Ans.** 320 beads

2 $90 \times 4 = 360,\ 30 \times 8 = 240,$
  $360 + 240 = 600$ **Ans.** 600 pieces

3 $5 \times 2 = 10,\ 3 \times 5 = 15,$
  $10 + 15 = 25$ **Ans.** $25

4 $15 \times 4 = 60,\ 12 \times 4 = 48,$
  $60 + 48 = 108$ **Ans.** 108 people

5 $80 \times 4 = 320,\ 40 \times 6 = 240,$
  $320 - 240 = 80$ **Ans.** 80 bagels

6 $5 \times 7 = 35,\ 5 \times 5 = 25,\ 35 - 25 = 10$
          Ans. 10 black stones

7 $32 \div 8 = 4,\ 48 \div 8 = 6,$
  $6 - 4 = 2$ **Ans.** 2 clips

8 $72 \div 9 = 8,\ 36 \div 6 = 6,$
  $8 - 6 = 2$ **Ans.** 2¢

29 Mixed Calculations pp 58,59

1 $10 \times 3 = 30$ **Ans.** 30 ft.

2 $10 \times 4 = 40$ **Ans.** 40 ft.

3 $10 \times 5 = 50$ **Ans.** 50 ft.

4 $7 - 1 = 6,\ 10 \times 6 = 60$ **Ans.** 60 ft.

5 $9 - 1 = 8,\ 8 \times 8 = 64$ **Ans.** 64 ft.

6 $10 \times 5 = 50$ **Ans.** 50 m

7 $10 \times 8 = 80$ **Ans.** 80 m

8 $15 \times 10 = 150$ **Ans.** 150 m

9 $3 \times 20 = 60$ **Ans.** 60 m

10 $12 \times 15 = 180$ **Ans.** 180 m

30 Mixed Calculations
pp 60,61

1) $100 \times 2 = 200$
$200 - 10 = 190$ **Ans.**   190 in.

2) $100 \times 2 = 200$
〔Also, $100 + 100 = 200$〕
$200 - 20 = 180$ **Ans.**   180 in.

3) $150 \times 2 = 300$
〔Also, $150 + 150 = 300$〕
$300 - 10 = 290$ **Ans.**   290 in.

4) $150 \times 2 = 300$
〔Also, $150 + 150 = 300$〕
$300 - 20 = 280$ **Ans.**   280 cm

5) $200 \times 2 = 400$
〔Also, $200 + 200 = 400$〕
$400 - 10 = 390$ **Ans.**   390 cm

6) $200 \times 2 = 400$
〔Also, $200 + 200 = 400$〕
$400 - 25 = 375$ **Ans.**   375 cm

31 Mixed Calculations
pp 62,63

1) $11 - 1 = 10,\ 10 \div 2 = 5$
     Ans.   5 white flowers

2) $12 - 2 = 10,\ 10 \div 2 = 5$
     Ans.   5 white flowers

3) $15 - 3 = 12,\ 12 \div 2 = 6$
     Ans.   6 white flowers

4) $24 - 6 = 18,\ 18 \div 2 = 9$
     Ans.   9 white gumballs

5) $30 - 14 = 16,\ 16 \div 2 = 8$ **Ans.**   8 walnuts

6) $11 + 1 = 12,\ 12 \div 2 = 6$ **Ans.**   6 boys

7) $12 + 2 = 14,\ 14 \div 2 = 7$ **Ans.**   7 boys

8) $15 + 3 = 18,\ 18 \div 2 = 9$ **Ans.**   9 boys

9) $14 + 2 = 16,\ 16 \div 2 = 8$
     Ans.   8 white shirts

10) $15 + 3 = 18,\ 18 \div 2 = 9$
     Ans.   9 dark chocolates

32 Graphs & Tables
pp 64,65

1) (1)

Favorite fruit

grapes	apple	melon	orange	pear	peach
III	IIII	HHH IIII	HHH II	I	I

(2)

Favorite fruit

Fruit	grapes	apple	melon	orange	the others	Total
Number of people	3	4	9	7	2	25

2) (1) I
(2) novel⋯14   comic book⋯10
photo book⋯9   history book⋯3
(3) 4 students

3) (1) 2 m
(2) June⋯18 m   Jim⋯34 m
Tom⋯29 m   Cathy⋯24 m
(3) 16 m
(4) 5 m

33 Graphs & Tables
pp 66,67

1) (1) (2) (3) (4)

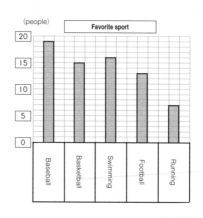

(5) Green

2) (1) (2) (3) (4)

34 Graphs & Tables pp 68,69

1 (1) A···64　B···61　C···55

(2)(3)(4)

Absent from Grade 3

Class \ Month	A	B	C	Total
April	25	21	19	65
May	13	10	17	40
June	18	19	9	46
July	8	11	10	29
Total	64	61	55	180

2 (1) 11 students　(2) 4 students

(3) ⓐ···36　ⓑ···35　ⓒ···37

(4) Class B

(5) ⓓ···36　ⓔ···34

　　ⓕ···26　ⓖ···12

(6) Novel

(7) 108

(8) Total students

35 Review pp 70,71

1 (1) $236 + 187 = 423$　**Ans.** 423 people

(2) $236 - 187 = 49$　**Ans.** 49 children

2 $450 m + 380 m = 830 m$　**Ans.** 830 m

3 $8 + 13 = 21$　**Ans.** 21 minutes

4 $260 g + 650 g = 910 g$　**Ans.** 910 g

5 $12 \times 15 = 180$　**Ans.** 180 blocks

6 $24 \times 5 = 120,\ 60 + 120 = 180$

　　Ans. 180 cm

7 $63 \div 9 = 7$　**Ans.** 7 boxes

8 $54 \div 8 = 6 \text{ R} 6$

Ans. 6 candies per child, 6 candies remain.

9 $63 \div 8 = 7 \text{ R} 7$

Ans. 7 bags, 7 oranges remain.

10 $22 - 4 = 18,\ 18 \div 2 = 9$　**Ans.** 9 girls

36 Review pp 72,73

1 $30 \times 25 = 750$　**Ans.** 750 flowers

2 $950 m - 670 m = 280 m$　**Ans.** 280 m

3 $2 \times 2 \times 3 = 12$　**Ans.** 12 L

4 $920 g - 350 g = 570 g$　**Ans.** 570 g

5 $75 \times 24 = 1,800$　**Ans.** 1,800 g

6 $85 \times 36 = 3,060$　**Ans.** 3,060 olives

7 $56 \div 8 = 7$　**Ans.** 7¢

8 $40 \div 6 = 6 \text{ R} 4$　**Ans.** 6 bunches

9 **Ans.** 8:15

10 $15 - 1 = 14,\ 6 \times 14 = 84$　**Ans.** 84 m

80 　ⓒ Kumon Publishing Co., Ltd.